International Conference on Smart Materials for Sustainable Construction in Civil Engineering (ICSSCE 2022)

IOP Conference Series: Materials Science and Engineering
Volume 1255

Bengaluru, India
8 - 9 July 2022

ISBN: 978-1-7138-6155-3
ISSN: 1757-8981

Printed from e-media with permission by:

Curran Associates, Inc.
57 Morehouse Lane
Red Hook, NY 12571

Some format issues inherent in the e-media version may also appear in this print version.

No changes have been made to the content of these proceedings. There may be changes to pagination and minor adjustments for aesthetics.

Printed with permission by Curran Associates, Inc. (2022)

For permission requests, please contact the Institute of Physics at the address below.

Institute of Physics
Dirac House, Temple Back
Bristol BS1 6BE UK

Phone: 44 1 17 929 7481
Fax: 44 1 17 920 0979

techtracking@iop.org

Additional copies of this publication are available from:

Curran Associates, Inc.
57 Morehouse Lane
Red Hook, NY 12571 USA
Phone: 845-758-0400
Fax: 845-758-2633
Email: curran@proceedings.com
Web: www.proceedings.com

TABLE OF CONTENTS

Preface

Due to Rapid Infrastructure development that is happening in and around, use of Smart and sustainable materials has become the need of the hour in construction industry. Use of smart materials in civil engineering is definitely a new approach for sustainable and durable construction of important structures. With minimal chance of load exceedance and collapse of structures, the time life of such structures can be increased with the use of smart materials. Therefore, it is very important to develop a smart system for concrete structures, which can minimize internal and external disturbances for structural safety and extension of its service life. The International Conference on **Smart Materials for Sustainable Construction in Civil Engineering (ICSSCE) 2022**, will show the way forward for the future of the construction industry. The conference will bring together the key players such as Academicians, Researchers, and Industry professional from the construction industry from with the country and around the globe. We hope that our Conference will offer a unique opportunity for interesting and fruitful interactions to all participants, providing them with memorable and pleasant experiences.

We take this opportunity to extend our sincere thanks to Hon'ble Chancellor, Vice-Chancellor, Pro- Vice Chancellor, Registrar and faculty and staff of Civil Engineering and the fraternity of REVA University for their guidance and support in organizing the conference. A conference of this kind and magnitude would not have been possible without the dedicated efforts of all people. I would like to express my gratitude to my colleagues, students and many others who are involved in the conference directly or indirectly for their wholehearted support and enthusiasm in organizing the conference. A Special Thanks to Dr. Krishna R. Reddy, Professor, University Scholar and Distinguished Researcher, University of Illinois, Chicago, USA, for working as Editor for this Conference.

IOP Conf. Series: Materials Science and Engineering 1255 (2022) 011001 doi:10.1088/1757-899X/1255/1/011001

Organizing Committee

Chief Patron

Dr. P. Shyama Raju Chancellor, REVA University

Patron

Dr. M. Dhanamjaya, Vice Chancellor, REVA University.

Dr. Rajashekhar C. Biradar, Pro-Vice Chancellor, REVA University.
Dr. N. Ramesh, Registrar, REVA University.

Honorary Chairs

Dr. D V S Bhagavanulu

Dean, IQAC & Academics, REVA University, India

Dr. B. P. Divakar

Dean R&I Council, REVA University, India

Dr. Sunilkumar S. Manvi

Dean, Engineering & Technology, REVA University, India

Chairman

Dr. Y Ramalinga Reddy

Professor & Director, School of Civil EngineeringREVA University, India

Conveners

Dr. S Srikanth Reddy

Associate Professor, School of Civil Engineering.

Dr. Kannam Praveen

Associate Professor, School of Civil Engineering.

Technical committee:

Dr. Mohd Zain Kangda, Associate Professor, School of Civil Engineering.

Peer Review Statement

All papers published in this volume have been reviewed through processes administered by the Editors. Reviews were conducted by expert referees to the professional and scientific standards expected of a proceedings journal published by IOP Publishing Publishing.

- **Type of peer review:** Double Anonymous
- **Conference submission management system:** We have formed a technical committee , headed by Dr. Mohd Zain Kangda, email id: mohdzain.kangda@reva.edu.in
- **Number of submissions received:** 25
- **Number of submissions sent for review:** 24
- **Number of submissions accepted:** 17
- **Acceptance Rate (Submissions Accepted / Submissions Received × 100):** 68
- **Average number of reviews per paper:** 2
- **Total number of reviewers involved:** 8
- **Contact person for queries:**
  Name: Dr. Kannam Praveen
  Email: kannam.praveen@reva.edu.in
  Affiliation: REVA University

Lateral gap analysis for Heterogeneous traffic

Siddharth Chogule [1], **Kanimozhee S** [2] **Madhu K A**[2]

1Post Graduate Student, School of Civil Engineering, REVA University, Bangalore, India
[2]Assistant Professor, School of Civil Engineering, REVA University, Bangalore, India
(Email:, r19mte17@ce.reva.edu.in, madhu.ka@reva.edu.in, kanimozhee.s@reva.edu.in)

Abstract: Indian traffic mainly composes of heterogeneous traffic with minimum or no lane discipline, due to this there are various vehicular interactions in the traffic stream. Lateral gap is the lateral distance between two vehicles flowing in the traffic stream. In this paper an attempt is made to analyse the safe gap maintained by the various category of vehicles with respect to their traffic characteristics and how the road width is affected by area occupancy. Methodology for collection of microscopic data is elaborated. Variation in lateral gap is observed due to the speed of adjacent moving vehicles, road width, and composition of vehicle. The relation between flow and area occupancy determines the utilisation of road width on the selected stretch.

Keywords: Area occupancy; Flow; Heterogeneous traffic; Lateral gap; Speed.

1. Introduction

In developing countries like India congestion problems are prevalent, affecting the road users due to increase in a number of vehicles. To combat the traffic congestion problem, an extensive understanding and analysis of the traffic stream behavior are necessary. Study of various traffic characteristics is necessary for planning, designing and operation of roadway facilities, in addition to regulation and control of traffic.

Traffic stream behavior is an outcome of the human driving process, which is a natural phenomenon. Thus, this behavior is complex. Moreover, one cannot perform experiments in the real-world traffic stream to understand or improve the traffic movement patterns. The situations that are of interest to traffic engineers cannot be reproduced in the laboratory (road, drivers and vehicles). Traffic simulation is a powerful and cost-efficient tool for road design and evaluation of traffic management methods. Simulating means imitating the roadway and traffic conditions planned or present in the real world. Modelling the driver behavior requires a detailed understanding of the vehicular interactions with the surrounding traffic using gaps maintained between vehicles, governing speeds in various traffic situations, reactions in the form of acceleration or deceleration or veering, or a combination of these behaviors. Hence microscopic modelling of traffic flow can well represent the dynamics of vehicular movements and its lateral and longitudinal interactions within a traffic stream.

The traffic in developing countries is chaotic and does not move in well-defined channels thus increasing interactions between vehicles. To analyze such a stream, a study of the effect of adjacent and leading vehicles on behavior of a driver in the traffic stream is necessary. This effect causes drivers to maintain certain longitudinal and lateral gaps within traffic stream. Each driver will respond to particular situations of traffic stream, differently based on his/her perception capability and aggressiveness. Simultaneous lateral and longitudinal interactions make the system complex. The traffic stream in developing countries is more complex to analyze than developed countries due to some of its peculiar features such as heterogeneous nature of vehicles and no lane discipline and. They are described below in detail.

2. Methodology

The methodology describes the method of traffic data collection for this study, Further it describes data-extraction from the field for speed, volume, lateral gap and area occupancy. Consider a test vehicle moving in a heterogeneous traffic with weak lane discipline. The driver of this vehicle needs to check movement of several vehicles- those ahead of test vehicle as well as those moving alongside test vehicle for safe traversing. These vehicles affect the movement and position of test vehicle as it is manoeuvred by the driver in this traffic stream. Such vehicles are called interacting vehicles, since they influence the position of test vehicle in the next instance. Apart from longitudinal interactions, vehicles also interact laterally with test vehicle, since they can move sidewise in this traffic stream whenever the gap is available between vehicles ahead. Thus, a driver needs to have attention to both- in the front as well as sidewise to safely traverse in such traffic stream. Thus, interactions governing a vehicle position in this stream can be classified as 'longitudinal interactions' and 'lateral interactions'. Their variation with speeds and positions of interacting vehicles, vehicle types, width of road etc., need to be studied in detail. Parameters influencing the behavior of following-vehicle under the impact of heterogeneous traffic stream may include speeds of interacting vehicles, their lateral and longitudinal positions and gaps between them, the vehicle types, driver behavior, traffic stream density and road geometry. It is assumed that external characteristics (such as land-use, climate, pavement conditions, etc) have negligible influence on behavior of the vehicles. Vehicles are assumed to travel within the road width independent of roadway features such as lane marking, median type and width, etc. Various inter-vehicular parameters arise when a test vehicle interacts with other vehicles, which influence behavior of the driver.

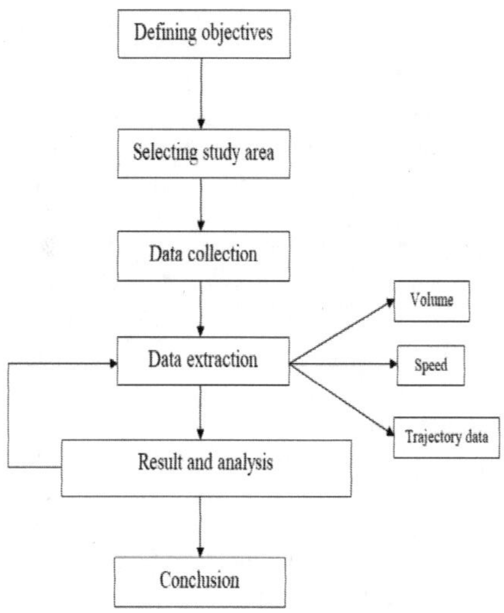

Figure 1: Schematic representation of methodology

2.1 Selection of study area

For the analysis of lateral gap and the occupancy of road mix-traffic, with vehicle composition of various classes is required to study the gap maintenance behavior for heterogeneous traffic. In this study, site selection is base on the these factors;

a) Road width

b) Intensity of traffic composition of all classes

c) Uninterrupted flow of traffic

d) Movement of vehicles in single direction.

The study was carried out in Bangalore, one of the major cities in India. The selected site is located near NH44, "Palanahalli, kattigenahalli, Bengaluru". The National Highway 44 connects the city centre and the International airport of Bangalore. It is the only means for commuters to reach their destination (fast). The site consists of two-lane service roads on both side of the national highway 44 of width 7.5m, and six-lane divided carriageway of 10.5m width. Speed limits on the service lane and the highway is 40kmph and 80kmph respectively. For the purpose of the study video-graphic survey had to be carried out, which required a foot over bridge to set up the camera for data collection.

2.2 Data collection
For the analysis of lateral gap various traffic characteristic are required such as speed, volume, flow, trajectory data etc. To collect all these parameters individually requires more man power and time consuming. In order to overcome this issue, video graphic survey was carried out on the selected site. The study area was divided into 2 sections depending on their road width and speed limits. Survey was carried out during the peak hours from 9am – 10am and 4pm – 5pm. Since the vehicle composition and interaction on the service lanes is heterogeneous and the vehicles utilize the road width, 4 hours of data was collected on the service lanes. On the highway the car and heavy vehicles move at higher speed which affects the light motor vehicles like two wheelers and three wheelers mostly occupy the shoulder lane which limits the interaction with other vehicle classes, this result in homogeneous traffic condition on the median lane and middle lane. Only two hours of data is collected for the highway section. A total of 6 hours of video graphic data was collected for extraction and analysis. Trap length of 20m for the trajectory data extraction was selected and marked for all the sections.

2.3 Data extraction
This is the curtail part of the study which requires accurate data for further analysis throughout the study. The data required for the analysis of lateral gap and area occupancy are explained below;

2.4 Speed and volume
The speed of interacting vehicle was calculated using "Traffic data extractor" (TDE) software. This software uses reference points on the video to mark reference lines for entry and exit of vehicle. Once a vehicle approaches the section, respective keys for vehicle class are entered when the vehicle travels through reference lines. The software is capable of 6 different vehicle class like Two-wheelers, Three wheelers, Car, LCV, Heavy and Multi-axle. The software outputs the file in excel format, which consist of vehicle class, start time and end time of each vehicle, frame number and speed of the vehicle in meter per sec. 60meter trap length was marked to find the speed of the vehicles. Similarly to find the volume a reference like is dropped on the video to count the number of vehicles during the period.

Figure 2: Speed and volume extraction using traffic data extractor

Table 1: Traffic data extractor output for speed and volume

Veh no.	Frame no.	Start time(s)	End time(s)	Veh type	Speed (Km/hr)
1	417	5.405	13.914	Car	7.05
2	354	7.04	11.812	Car	12.57
3	495	8.308	16.516	Car	7.31
4	549	11.812	18.318	Car	9.22
5	533	12.88	17.784	Bike	12.23
6	811	14.882	27.06	Bike	4.93
7	632	13.58	21.088	Auto	7.99

2.5 Trajectory data
To find lateral gap between interacting vehicles in the stream for the collected video graphic data, a grid was plotted

on the trap length of 20m with 2m offset using "AutoCAD", for both 7.5m service lane and 10.5m National highway. The grid was then overlaid on the video using "MAGIX Movie edit pro" video editing software, approximately on the trap length marking. The software is able to run the video at 1 frame per sec which helps to measure the lateral gap between the interacting vehicles. When a two vehicle interact in the grid, an image is captured in MAGIX and imported back into AutoCAD. The distance is measured between the vehicles for the nearby interacting grid lines. The measured distance is then multiplied with the scale ratio for each of the grid line, which is the ratio of screen distance to ground distance. By multiplying the scale ratio to the measured distance, the actual distance maintained by the interacting vehicle can be determined. Similar process is obtained for all interacting vehicle and tabulated in the excel sheet with respect to the vehicle class.

Figure 3: Overlay of grid using MAGIX edit pro

Figure 4: Measuring lateral distance using AutoCAD

2.6 Area occupancy

It is defined as the number of vehicles occupancy a unit length of road at an instantaneous time. The density can be replaced with area occupancy as it is possible to identify the vehicle class occupying the road at any time period. The method used by Ch. Mallikarjuna et al., (2006) to calculate area occupancy is adopted. The formula is simplified in Raunak Mishra et al., (2017) is used in this study.

$$\textbf{\textit{Area occupancy}}\ (\%) = \frac{\sum_{1}^{i} t_i a_i}{TA}$$

t_i = Occupancy time for i vehicle(sec)
a_i = Area of the vehicle i occupying the road space (m^2)
T = Total observation period (sec)
A = Area of detection zone (m^2)

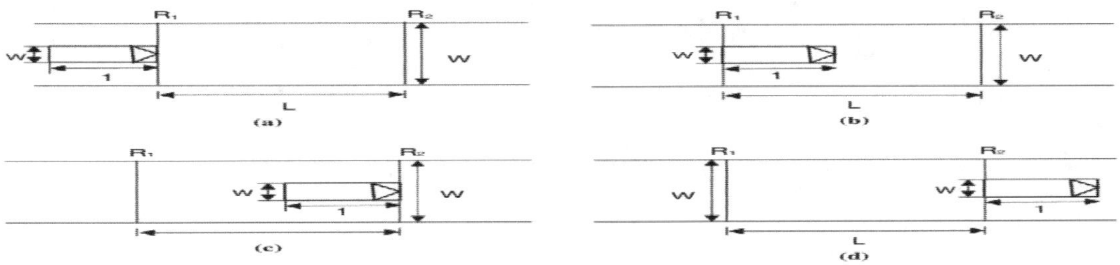

Figure 5: Schematic representation to determine area occupancy

In Figure 5, consider a stretch of road of length "L" and width "W". Two reference line for entry and exit R1 and R2. The time taken by vehicle a1 to travel the distance of its own length is taken as t1. When the rear bumper leaves R1 till the front bumper touches R2, the time taken to travel is noted as t2. The difference between t1 and t2 gives the occupancy time for the vehicle a1. Similarly, the area occupancy is calculated for all interacting vehicle.

3. Analysis

The extracted data from the methodology is used for analysis, to understand the relationship between various traffic characteristics such as speed, volume, lateral gap, road width usage and area-occupancy. The extracted data is difficult to analyze with no proper representation. The relations are plotted using R-programming language in "R-Studio". A graphical representation for all vehicle classes depending on their physical characteristics are been plotted;

a) Speed and lateral gap.
b) Speed and volume.
c) Area-occupancy and lateral gap.
d) Area-occupancy and average speed.
e) Road width usage and area-occupancy.
f) Road width usage and lateral gap.

3.1 Speed and lateral gap

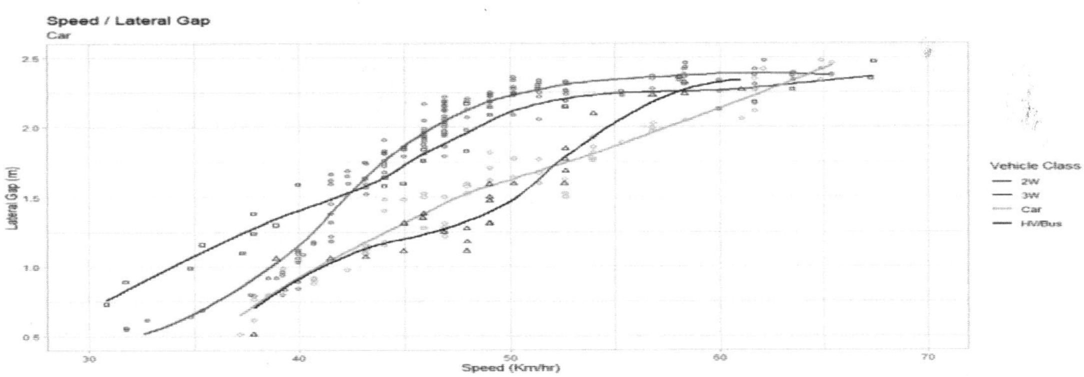

Figure 6: Speed / Lateral Gap for all vehicle classes with respect to car on 7.5m wide road.

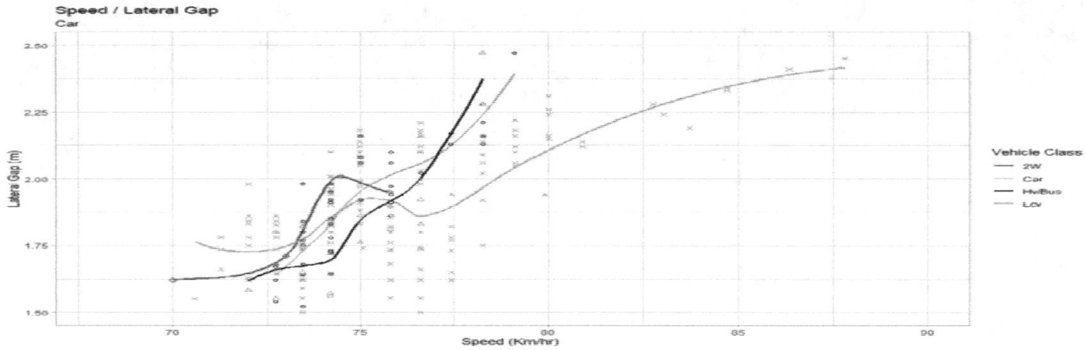

Figure 7: Speed / Lateral Gap for all vehicle classes with respect to car on 10.5m wide road.

In Figure 6 and figure 7, a non-linear between speed and lateral gap is observed. In both the section (7.5m and 10.5m wide road) it is observed that, at lower speed car maintains lesser gaps and increases with increase in speed.

3.2 Speed and volume

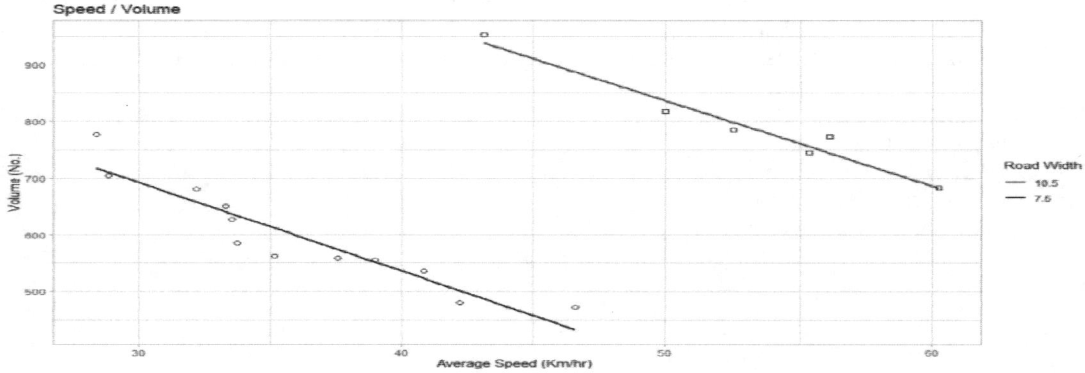

Figure 8: Speed / Volume for both 7.5m and 10.5m wide road.

The speed and volume graph shows linear relations between speed and volume for both the sections. It is observed that, with increase in volume the speed of vehicle reduces. From previous relations where the lateral gap is less at lower speeds, the number of interaction increases as volume increases.

3.3 Area-occupancy and lateral gap

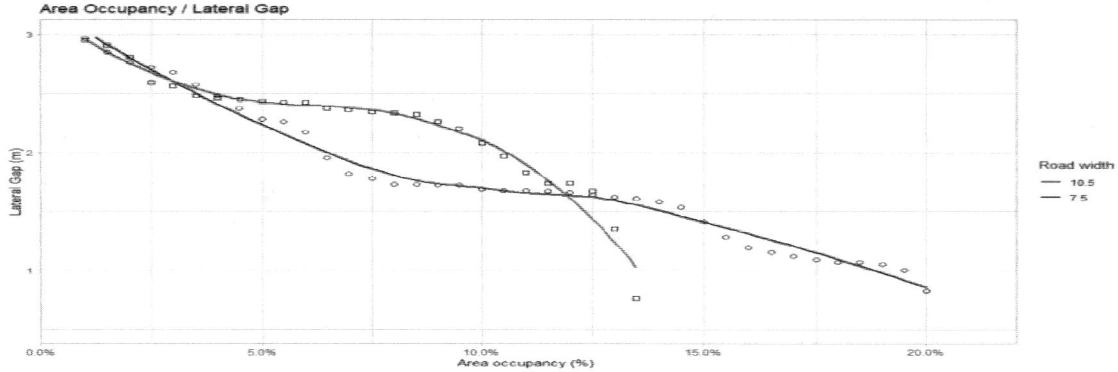

Figure 9: Area-occupancy / Lateral Gap for 7.5m and 10.5m wide road.

From Figure 9, it is observed that lateral distance between the subject vehicles is more when the occupancy is less and shortens when the occupancy is more. The area occupancy for 7.5m wide road is more compared to 10.5m wide road.

3.4 Area-occupancy and Average speed

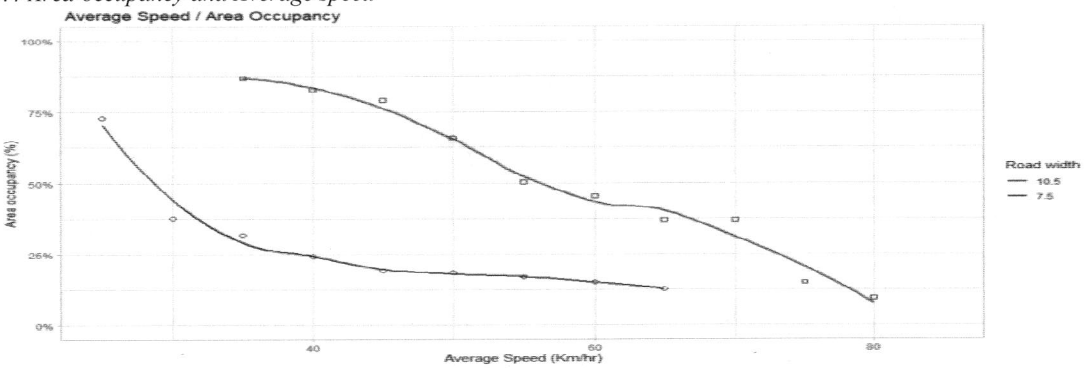

Figure 10: Area-occupancy / Average speed for 7.5m and 10.5m wide road.

In Figure 10, the relation between area-occupancy and average speed is non-linear which indicates that area-occupancy percentage is more at lower speed and reduces as the speed increases.

3.5 Road width usage and Area-occupancy

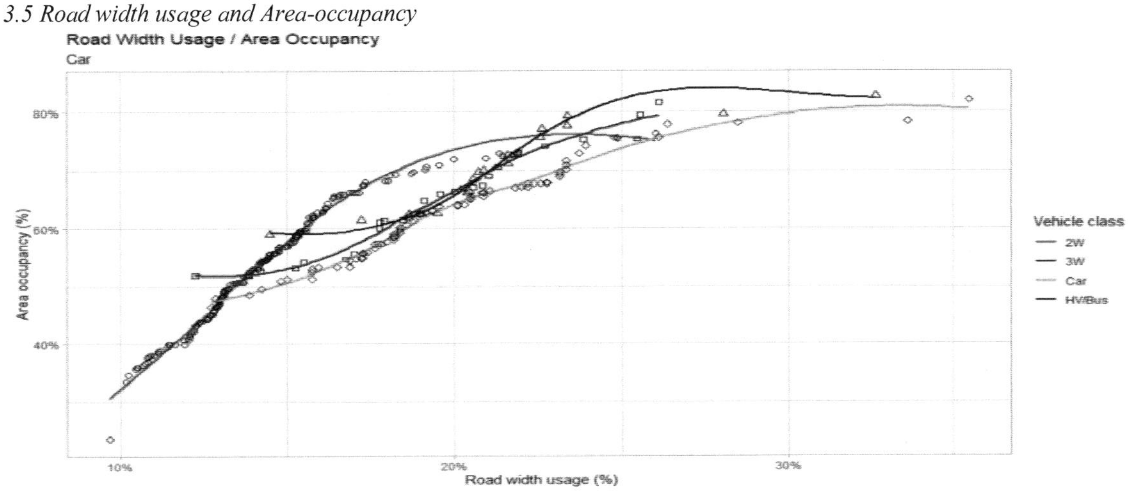

Figure 11: Road width usage / Area-occupancy for all vehicle class on 7.5m wide road.

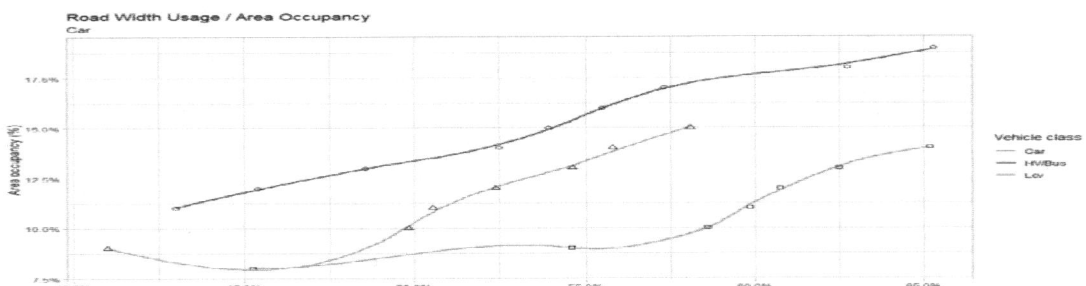

Figure 12: Road width usage / Area-occupancy for all vehicle class on 10.5m wide road.

Road width usage (%) depend upon the width of the vehicle class interacting in the stream and the gap maintained by them on the width of the road. From Figure 11 and Figure 12, the road width usage is more with respect to car when the area-occupancy increases. At the maximum area-occupancy in free flow for 7.5m wide road the road usage is constant. On 10.5m wide road the area-occupancy is less due to higher speed and larger width of road.

IOP Conf. Series: Materials Science and Engineering 1255 (2022) 012001 doi:10.1088/1757-899X/1255/1/012001

3.6 Road width usage and Lateral Gap

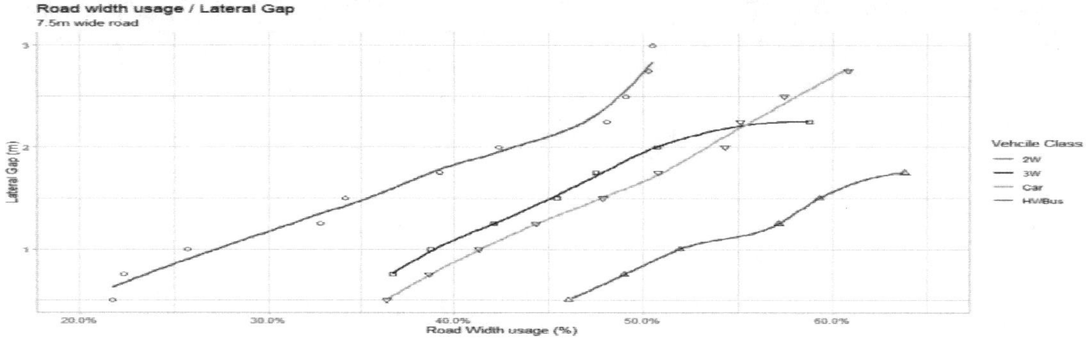

Figure 13: Road width usage / Lateral Gap for 7.5m wide road.

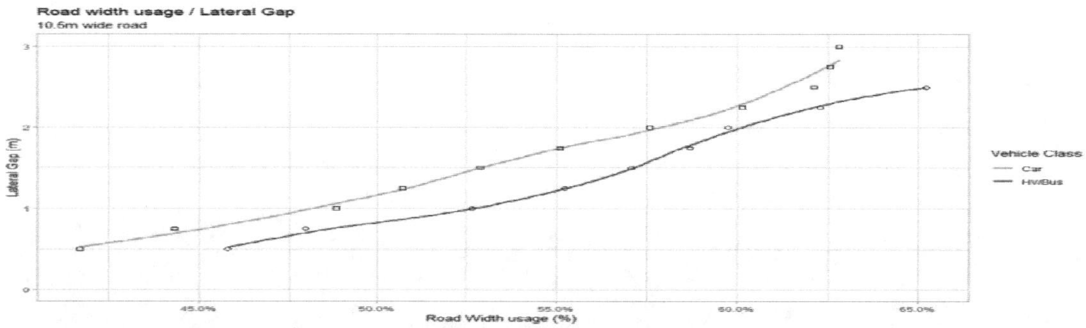

Figure 14: Road width usage / Lateral Gap for 10.5m wide road.

For different vehicle classes the average gap maintained by the vehicles on the different road width show that, as the gap increases between the interacting vehicles the road width usage also increases as the vehicles tent to occupy the available space along the width.

4. Results

From the relations generated for various traffic characteristic in heterogeneous traffic, the interaction for most of the vehicle classes have been analyzed to understand the effect of each characteristic over other. The variation in lateral gap acceptance depends on speed, type of vehicle, area occupancy and road width. The maximum and minimum lateral gap for car with respect to speed is 0.48m and 2.42m on 7.5m wide road and 1.5m and 2.46 on 10.5m wide road. The maximum volume of two-wheelers form 4hours of data were 3898 number of vehicles on 7.5m wide road and maximum volume of 2800 number of cars on 10.5m wide road for two 2hours of data. The area-occupancy for 10.5m wide road is less than 20% due to larger width of road and higher design speed and 83% for 7.5m wide road. The road width usage is 65% on 10.5m wide road and 64% on 7.5m wide road.

Discussion

In this study, an attempt is made to study the lateral interaction of vehicles in mix traffic condition. An innovative method is developed for measuring the lateral distance between vehicles accurately. A detailed methodology for manual data extraction is explained, with the help of various software used for image processing. R-programming language is used to analyze the extracted data. From the results obtained by graphical representation for the correlation between traffic characteristic affecting the lateral gap maintaining behavior of different vehicle classes. It is observed that with change is speed of the subject vehicle interacting with the adjacent vehicle, the gap maintained is more with respect to road width. The gap maintenance varies due to various physical characteristics of vehicle and their dynamic behavior during maneuvering in the traffic stream. It is observed that the smaller sized vehicles try to navigate through the available gap which increase the cause of accidents and affects the speed of adjacent moving vehicle. The results show similar relations compared to the literature review for the modeled values of gap acceptance behavior and area occupancy.

Conclusion

Lateral gaps are evaluated between various pairs of vehicles as they move in the traffic stream, and their variation with the speeds of vehicles, area-occupancy, and road width as per the objective.

1. The lateral gap varies for different vehicle interaction with respect to the speed of subject vehicle and adjacent vehicle.
2. It is observed that vehicles maintain lesser lateral gap between similar interacting vehicle types than other vehicle types. Due to this behavior, lateral gap cannot be assumed as the sum of individual contribution of vehicles constituting the gap.
3. The area-occupancy of each vehicle varies depending on their physical dimensions and the period required to travel through reference points. When two or more vehicles interact the area-occupancy increases which restrict the vehicles to maintain smaller gaps.
4. The increase in road width usage is due to the varying widths of the vehicles and their gap maintaining behavior.

References

[1] Anuj Kishor Budhkar, Akhilesh Kumar Maurya. “Characteristics of lateral vehicular interactions in heterogeneous traffic with weak lane discipline”. J. Mod. Transport. (2017) 25(2):74–89.
[2] V. S. H. Bangarraju, K. V. R. Ravishankar, Tom V. Mathew. “Analysis of lateral distance keeping behaviour in mixed traffic conditions with little lane discipline”. International Journal for Traffic and Transport Engineering (2016), 6(4): 431 – 443.
[3] C. Mallikarjuna, Budde Tharun, Dibyendu Pal. “Analysis of the Lateral Gap Maintaining Behaviour of Vehicles in Heterogeneous Traffic Stream”. Procedia - Social and Behavioral Sciences 104 (2013) 370 – 379
[4] Kavitha Madhu, R Sivanandan and Karthik K Srinivasan. “Identification of different vehicle-following manoeuvres for heterogeneous weak-lane disciplined traffic condition from vehicle trajectory data”. IOP Conf. Series: Earth and Environmental Science 491 (2020) 012052.
[5] Dibyendu Pala, C. Mallikarjuna. “Analysis of the Effect of Variable Lateral Gap Maintaining Behaviour of Vehicles on Traffic Flow Modeling”. Procedia Engineering 142 (2016) 198 – 204
[6] Raunak Mishra, Pallav Kumar, Shriniwas S. “Novel Area Occupancy–Based Method for Passenger Car Unit Estimation on Multilane Urban Roads Under Heterogeneous Traffic Scenario”. Journal of the Transportation Research Board, No. 2615, 2017, pp. 82– 94.
[7] Geetimukta Mahapatra, Akhilesh Kumar Maurya. “Study of vehicles lateral movement in non-lane discipline traffic stream on a straight road”. Procedia - Social and Behavioural Sciences 104 (2013) 352 – 359
[8] Sai Kiran, Ashish Verma. “Review of Studies on Mixed Traffic Flow: Perspective of Developing Economies”. Springer International Publishing Switzerland.
[9] Banihan Gunay. “Car following theory with lateral discomfort”. Elsevier Transportation Research Part B 41 722–735.
[10] Syed Omar Ballari. “Area Occupancy Characteristics in Traffic Flow on Urban Highway: A Case Study”. Journal of Advanced Research in Dynamical and Control Systems (2019)
[11] O.P.Korobeynikov,and.A.Trifilova.“TheIntegrationoftheStrategicandInnovationManagement”, Management in Russia and Abroad, vol.4, 2001.URL:www.cfin.ru/press/management/.(27.11.2015).
[12] E.V.Tolkacheva. “Strategic Controlling in the Enterprise Management System”, Management in Russia and Abroad, vol.4, pp.109-118, 2004.
[13] S.M.Rezer.“Fundamentals of modeling optimal logical systems delivery, Transport Innovations”, *Scientific and Technical Journal*, vol.**3** (18), 2014.

Crack Healing in concrete using Bacteria

Prashanth N[1], Dr Y Ramalinga Reddy[2], Karuna P Rao[3]
[1] *Assistant Professor, School of Civil Engineering, REVA University, Bangalore.*
[2] *Director, School of Civil Engineering, REVA University, Bangalore.*
[3] *UG Student, School of Civil Engineering, REVA University, Bangalore*

Abstract. One of the most often utilized building materials is concrete, yet it is prone to cracking. The structure is far more prone to fail due to cracks. There are several methods for, but they are not permanent and are not environmentally friendly. The inclusion of microorganisms in concrete can be a long-term option for crack healing or filling. It also influences the compressive and split tensile strength and maintain strength after crack formation to certain amount. Calcium Carbonate (Lime) production prevents cracks from recurring in the same location and serves as a permanent and long-term treatment. From the results obtained, it could be concluded that the product is more durable and requires less maintenance.

1. Introduction

Even though concrete is largely used it has many drawbacks. These drawbacks are a concern. Finding an appropriate repair procedure for restoring concrete strength is therefore critical.
Hence, these biological concepts are applied to concrete, an innovative material called Bacterial Concrete could be created. In this type, the concrete is induced with a particular type of bacteria. The calcium carbonate precipitates as a result of a few chemical processes, filling in the cracks when it does so.

2. Experimental Procedure

Based on the research done and the data collected, it was observed that the bacteria from the 'Bascilius' family were used. We have used is available in powder form which is known as "Bacilius subtilus".
According to the Indian standards code book, all the materials used are tested for different physical parameters. The grade of concrete chosen is M20 grade. The cubes are casted in two ways i.e., bacteria induced concrete cube and the conventional concrete cube.
In accordance with procedure specified in the code book the specimens are tested. The values are noted down. The bacteria induced cubes are left to self-heal whereas the conventional cubes are injected with the bacteria and adequate moisture is provided. Further, the cubes are again tested for compressive strength.

2.1. Selection of Bacteria

Bacteria such as Bacillus Subtilis, Bacillus Pasteurii, Bacillus Cohnii, Bacillus Licheniformis, and others can be included in concrete. Bacillus Subtilis was chosen because it produces calcium carbonate and because it was readily available at AIKTC's pharmacy department.
Preparation of the sample:
The powder form of bacteria procured is mixed into an agar solution and maintained at 30°C for 48 hours. Bascilius subtilius strains are detected under an electron microscope after 48 hours. The concentration of bacteria is 3 x 10^{10} gm/cm^3. The colour is pink and the cell walls are negatively charged.

2.2. Preparation of Concrete Mix

We used M20 concrete grade in our experiment. The ratio as per IS 456 for M20 grade is 1:1.5:3.
Quantities of ingredients required for M20 mix of (15 x 15 x 15 cm^3)
Cement (1) = 1.44kg
Fine aggregate (1.5) = 1.77kg
Coarse aggregate (3) = 3.55kg
Water Cement ratio = 0.5
Quantity of Bacteria = 200gm (powder form with 3 x 10^5 cells/cm^3)

2.3. Method of Mixing Bacteria into to the Concrete

In our study, we employed the direct method, which involved first sterilising the measuring jars for five minutes at a temperature of roughly 1000°C. After it has slightly cooled, the bacterial solution is put in to the measuring jar. The flask is placed beneath the candle to make sure it is not contaminated by outside microorganisms before the solution is transferred into the jar. Before using the water for concrete immersion, the bacterial solution has been properly incorporated into it.

2.4. Compressive Strength Value (Pre-Cracked Value)

After their respective days of curing, the concrete cubes are removed from the tank. Under laboratory conditions, the cubes are allowed to dry. After the cube had dried completely, it was kept for testing in the compressive testing machine. All sample specimens were tested. Right until the specimen breaks, the force is applied gradually, without shock, at a rate 5.2 KN/sec. The maximum load as well as any unusual failure type characteristics are noted.

Bacterial Reaction:

The sequence of events that lead to the formation of calcium carbonate

$$Ca\,(NH_2)_2 + H_2O \rightarrow NH_2\,CaO + NH_3$$

$$NH_2\,CaOH + H_2O \rightarrow NH_3 + H_2\,CO_3$$

$$H_2\,CO_3 \leftrightarrow HCO_3- + H^+$$

$$2NH_3 + 2H_2O \leftrightarrow 2NH_4+ + 2OH^-$$

$$Ca^{+2} + \text{cell} \rightarrow \text{cell - } Ca^{+2}$$

$$\text{Cell-}Ca^{+2} + CO^{2-} \rightarrow \text{cell-}CaCO \downarrow$$

2.5. Compression testing Strength (After Healing):

When the bacterial and treated conventional cubes', cracks have self-healed and healed respectively. The cubes which are cured for 28 days are used for testing the compressive strength.

3. Results and Discussion

The obtained results of the compressive strength of the tested specimens are tabulate below:

Table 1.1. Compressive strength of concrete afte

	Load	Compressive Strength	Increase in strength
Conventional 1	395 kN	17.5 mpa	--
Conventional 2	390 kN	17.33 mpa	--
Bacterial	410 kN	18.22 mpa	4.83 %

Table 1.2. Compressive strength of concrete afte

	Load	Compressive Strength	Increase in strength
Conventional 1	482 kN	21.42 mpa	_
Conventional 2	480 kN	21.3 mpa	_
Bacterial	555 kN	30.1 mpa	13.34 %

Table 1.3. Compressive strength of concrete after 14 days

	Load	Compressive Strength	Increase in strength
Conventional 1	630 kN	28 mpa	_
Conventional 2	610 kN	27.1 mpa	_
Bacterial	720 kN	32 mpa	13.89 %

Table 1.4. Compressive strength of concrete after

	Load	Compressive Strength	Increase in strength
Conventional 1	770 kN	34.22 mpa	_
Conventional 2	700 kN	32.0 mpa	_
Bacterial	880 kN	39.44 mpa	16.48 %

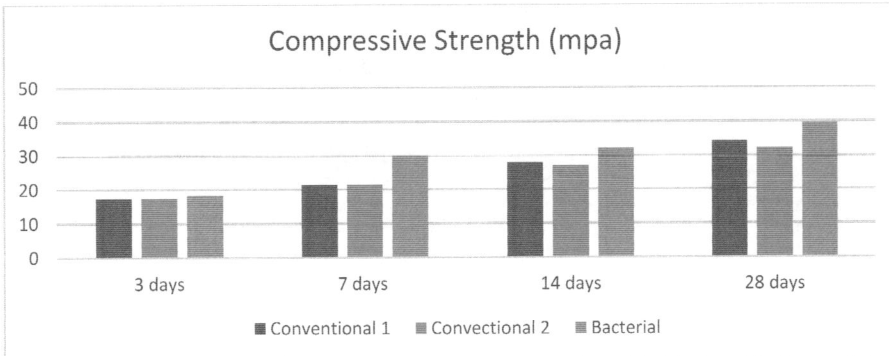

Graph 1.1 This graph shows the comparison between compressive strengths of all the specimens tested.

3.1 Healing of Cracks:

Concrete cracks are cured by injecting bacteria, which then self-heal because of the bacterial reaction (period 2 to 3months). Below is an electronic microscopic image of the hydrated structure of cement concrete and cement concrete with bacterial cell concentration.

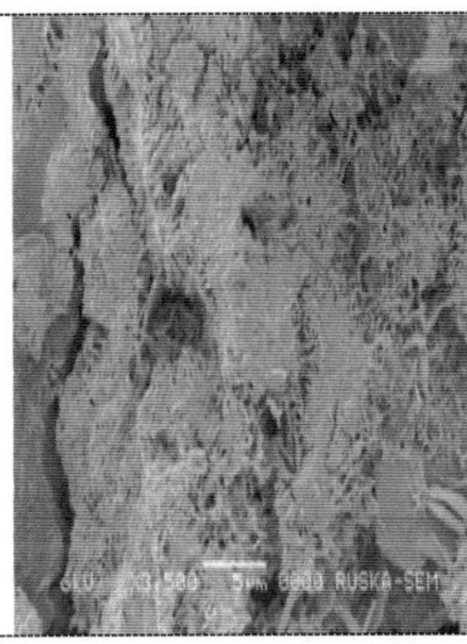

Figure 1. The Cell concentration is nil

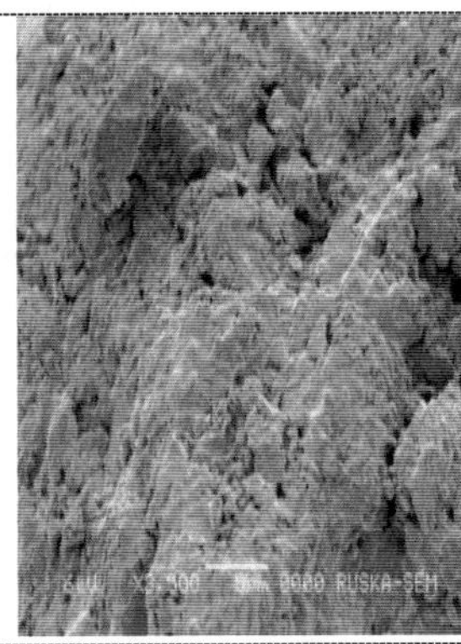

Figure 2. The Cell concentration is $3x10^5$ cells/cm^3

3.2 *Compressive Strength (after healing)*

The obtained results of the compressive strength of the tested specimens after healing are tabulated below:

Table 1.5. Compressive strength of self-healed concrete (y

	Load	Compressive Strength	Increase in strength
Conventional 1	770 kN	34.22 mpa	–
Conventional 2	700 kN	32.0 mpa	–
Bacterial	880 kN	39.44 mpa	16.48%

Table 1.6. Compressive strength of healed concrete cub

	Load	Compressive Strength	Increase in strength
Conventional 1	820 kN	36.44 mpa	6.91 %
Conventional 2	752 kN	34.66 mpa	6.92 %
Bacterial	978 kN	42.66 mpa	10.1 %

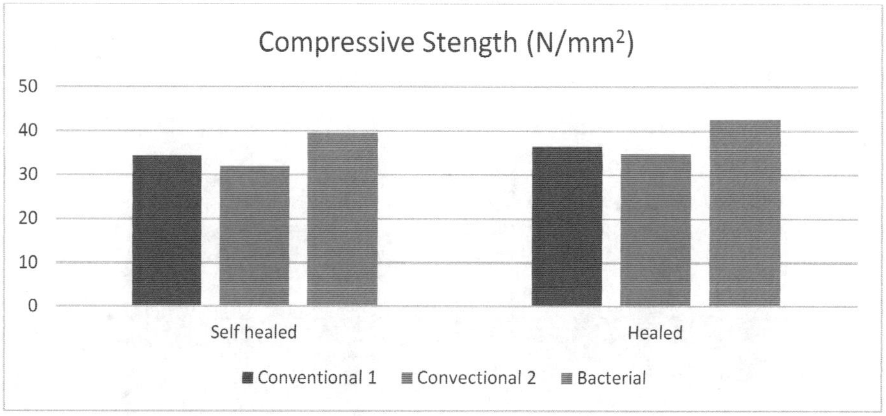

Graph 1.2 Compressive strength of self-healed crack is more in comparison with treated crack by around 4- 5%

4. **Conclusion:**

The greater the amount of $CaCO_3$ precipitation, the higher the self-healing effect. The 28-day curing cube shows the biggest improvement among all ages, with a strength gain of 16.48 %. There is 10.1% gain in strength of the self-healed cube. There is increase in of strength by 6% -6.5 % in the healed cubes. Cracks width of more than 0.8mm do not repair on their own. Cracks having width more than 0.6mm are not cured among those treated. As a result of the above, it may be concluded that self-healing bacterial concrete can perform better than ordinary concrete.

5. References

[1] Ramin Andalib, Muhd Zaimi Abd Majid, Mohd Warid Hussin, Mohanadoss Ponraj , Ali Keyvanfar,Jahangir Mirza, Han-Seung Lee ''Optimum concentration of Bacillus megaterium for strengthening structural concrete'' (2016).. Elsevier
[2] M. V. Seshagiri Rao, V. Srinivasa Reddy Ch. Sasikala ''Performance of Microbial Concrete Developed Using Bacillus Subtilus JC3'' (2017). Elsevier.
[3] Kunamineni Vijay, Meena Murmu, Shirish V. Deo. ''Bacteria based self-healing concrete – A review'' (2017). Elsevier.
[4] Jianyun Wang & Henk M. Jonkers & Nico Boon & Nele De Belie. ''Bacillus sphaericus LMG 22257 is physiologically suitable for self-healing concrete'' (2017). Springer.

[5] Nguyen Ngoc Tri Huynh, Nghai Mai Phuong, guyen Phung Anh Toan, guyen Khanh Son ''Bacillus subtilis HU58 immobilized in micro pores of diatomite for using in self- healing concrete'' (2017).. Elsevier.
[6] Ms. B.Arthi, Ms. K. K. Dhaarani (2016). ''A study on strength and self-healing Characteristics of Bacterial concrete'' IJETT.
[7] Wasim Khaliq, Muhammad Basit Ehsan (2016). ''Crack healing in concrete using various bio influenced self-healing Techniques'' Elsevier.
[8] Kim Van Tittelboom, Nele DeBelie, Willem De Muynck, Willy Verstraete (2010). ''Use of bacteria to repair cracks in concrete'' Elsevier
[9] Henk M. Jonkers, Arjan Thijssen, Gerard Muyze, Erik Schlangen (2010). ''Application of bacteria as self-healing agent for the development of sustainable concrete'' Elsevier.

IOP Conf. Series: Materials Science and Engineering 1255 (2022) 012003 doi:10.1088/1757-899X/1255/1/012003

Experimental Investigation of Aluminium Dross and GGBS in the Production of Eco-Friendly Concrete

Arpitha D J[1*], Dr. Kannam Praveen[2]

1,Research Scholar, 2 Associate Professor, School of Civil Engineering, REVA University, Bengaluru – 560064, Karnataka, India. (R21PCV05@reva.edu.in, kannam.praveen@reva.edu.in)

Abstract. The use of alternative materials in production of concrete is being well normal, since it makes two possibilities in improving within the concrete composites, additionally except standard economic system. This paper offers laboratory research record approximately the utilization of aluminium dross (AD), floor granulated blast furnace slag (GGBS) as partial alternative of cement. The partial replacement of cement turned into performed in specific trail blend for M35 grade of concrete one by one. The blend trail with distinct five percent of Aluminium dross and GGBS through zero%, 10%, 20%, 30%, forty% so that you can discover the most reliable percent use in concrete. From those trial mixes, percentage optimization of Aluminium dross and GGBS utilized in concrete for the in addition investigation. It became discovered that five% of Alluminium Dross and, 30% replacement of GGBS respectively offers most strength to concrete.

1.Introduction

In present days, civil engineering sector cannot be considered without concrete, it's miles the final task to bring the most results at affordable input without any impact to the purpose of work but environment. Concrete specially the traditional concrete is the use of global within the construction streem particularly with the material like water and additive cement material, conventional silicacalcareous based aggregates (sand and gravel) and chemical admixtures (extraordinary plasticizers, air entertainers, retarders and so forth.). Although regarding natural notable aggregates are superior to every other fabric in producing the concrete, their orign is continuously being depleted by the criteria of overexploitation in the course of the global criteria.In context to the problem partial or complete opportunity of notable aggregates by the alternative well matched substances like sintered based fly ash, Crushed type rock dust, iron dust, glassy type powder, recycled concrete dirt, and others are triying to researched in steem of keeping the stabilised pattern of ecological concept.

Aluminium dross AD and GGBS are the waste products at some stage in the manufacture of aluminium and Iron respectively. Those can efficiently use within the concrete production as alternative with sand and cement. This may additionally moreover result the trade in physical and chemical residences of concrete like compressive electricity, tensile power, flexure electricity.

Aluminium dross is a product obtained from the manufacturing process of Aluminium. Lots of strength is fed on it to get higher Al from the dross; thus by these process electricity that may be reserved if the dross is processed and implemented as an serviced engineering cloth.

Similarly, GGBS is a product which is received from the iron making industries as the growing of industrialization and urbanization is extra developing the producing of iron is extra and acquiring by way of products from the industries is also extra. By making use of this by products in concrete as partial alternative of OPC may be very beneficial in developing the excessive best concrete.

2.Concept of Eco-Friendly Concrete

Construction materials are actually judged through their ecological traits too. The use of business waste consisting of fly ash mineral , GGBS mineral, silica based mineral in concrete and other building substances is gaining importance daily due to several blessings similarly to the discount in fee (Erdoğdu&Türker, 1998; Ganesh Babu&Sree Rama Kumar, 2000; Mazloom,Ramezanianpour, & Brooks, 2004; Oner, Akyuz, &Yildiz, 2005). Various researches have attempted fly ash, floor GGBS, rice husk ash, and silica fume mineral as an attempt of partial alternative of cement in traditional and structuralstremgth concrete. The mining activity of sand has spoiled the streem ecology of almost all different riverchannels and flood plains. The network has being triedto find out options to naturally obtained river sand for process of construction. Some of the alternatives being considered encompass crushing hard rock to fines (synthetic sand), recycled excellent combination, mining for sand in submerged areas, and generating sand from molten iron slag which is a derivative of iron and metallic enterprise. About nearly 30 per cent of the cast iron ore being utilized in steel manufacturing turns to slag and this could be applied to meet the existing-day sand deficiency (Manning &Fruehan, 2001). Slag sand is long lasting, has consistent shape and volume, and is safe for use in production. The slag sand is more superiorthan a river sand due to the fact the latter consists of fossils and different abnormal debris which are silted particles and clay which impacts the best and sturdiness of concrete. In an effort to protect the natural resources and to eliminate the disposal problems of demolished concrete, which is readily available, concrete recycling has gained importance.

3. Materials

Material characterization is a necessary step which is to be conducted before conducting any experimental work as per IS code provisions. This study includes the materials which are cement, herbal sand – first-rate mixture, beaten rock - coarse aggregate, water, similarly to these aluminium dross and GGBS, a replacement to cement and quarry dust a substitute to sand are used. This material characterization helps to an engineer during design of concrete mixes of required strength. Brief details regarding materials and their properties are discussed below.

3.1Portland cement

Cement is a binding fabric which binds the constituents which include quality aggregates and coarse aggregates within the presence of water. Cement is the most vital detail inside the concrete. It is manufactured through a meticulously based chemical blending of calcium, aluminium, silicon, iron and other factors. Generally, the materials employed for the manufacture of cement includes a main amount of limestone, chalk, shells, marl fused with shale mineral, slate, clay based mineral , industrial blast furnace slag and iron ore. These elements are treated for the heating procedure at multiplied temperature which produces a rock-like cloth that's then floor/beaten into great residue that we universally name it as cement

IOP Conf. Series: Materials Science and Engineering 1255 (2022) 012003 doi:10.1088/1757-899X/1255/1/012003

3.2Aggregates

The aggregates will cowl a chief element in the matrix of concrete. The homes of aggregates will play a prime position inside the power gaining of the concrete. Aggregates are of two kinds specifically coarse and nice aggregates.

3.3 Coarse Aggregate

Stones obtained from Crushed quarry are normally treated as coarse mixture (CA) in concrete which should own its characteristics along with more dense characteristics, long lasting, well graded and higher surface for robust attachment with the matrix of mortar. The coarse aggregates have to be relatively cubical, spherical and angular in form; flaky and elongated portions have to be avoided.

Coarse aggregates are those which are having their size greater than 4.75mm. Crushed rock of size 20mm down were used inside the gift take a look at.

3.4Fine Aggregate

The great combination hired inside the present look at turned into herbal sand received from nearby resources and was conforming to the grading of quarter-II of Table three of IS: 10262-2009. Its shape became a cubical in nature or completely rounded incharacterhaving a smoothy floor texture, there by a very well workability of casted concrete.Fine aggregates are those which skip thru 4.75mm IS sieve.

3.5Aluminium Dross

Disposal and recycling of dross is a international problem. The major mass of aluminium based dross is backfills in websites of landfill, which is anticipated to cause leachingeffect of noxious metal ions into the shallow floor water desk triggering critical pollution troubles. Additionally, if aluminium dross interacts with floor water, it releases harmful gases together with NH3, PH3, CH4, H2, and H2S. Aluminium based dross is a risky waste product which is received fromaluminium smelting system. The undertaking posed through aluminium dross to the surroundings has attracted the eye of researchers over time. Efforts are geared closer to setting this risky waste to efficient use through engineering applications. In the existing take a look at aluminium dross is mixed in concrete as a partial quantity substation of cement material raging by various probabilities.

3.6Ground Granulated Blast Furnace Slag (GGBS)

The GGBS mineral slag is a derivative of iron manufacturing employer. Iron ore material and lime content stone are fed into the furnacehooper and molten state slag material floats above the molten iron compound at a maintained temperature of about 1500°C to 1600°C as a resultant. The composition of molten slag approximately 30% to forty% Sio_2 and approximately 40% CaO, which is sort of equal to the chemical reactive composition material of Portland type of cement. The molten type of slag, is water quenched unexpectedly, after the molten iron is trapped off which ends in a glassy natured granulate. The molten slag includes especially siliceous based and alumininm based fills. For the spricaltextured glassy granulates fine grounding is reqired.

Table1: Physical characteristics of GGBS

Properties	**Values**
S G	2 .7
B D	1800 k g /m^3
MDD (kN/m^2)	2.56 gm/cc
Optimum moisture content (%)	20%
CBR values soaked and cured	32%

4. Methodology

In the present study the mixes have been prepared by using IS 10262-2019. Concrete mix with 35 grade of concrete is prepared. The concrete with water cement ratio of 0.4 has been prepared by considering a slump value of 100mm. In each trial the percentage of GGBS is varied at a rate of 10, 20, 30 & 40 percentages. The control mix (M1) without admixtures has been prepared.

Table 2: Mix Proportions

Mix	%Cement	%Aluminium dross	%GGBS	%Sand	%CA
M1	100	0	0	100	100
M2	85	5	10	100	100
M3	75	5	20	100	100
M4	65	5	30	100	100
M5	55	5	40	100	100

5. Results:

Table 3: Results

Mix	CS (MPa)		S T S (M Pa)		FS(MPa)	
	7 days	28 days	7 days	28 days	7 days	28 days
M1	35.36	51.24	2.36	3.42	6.72	9.74
M2	30.94	44.84	2.06	2.99	5.88	8.52

M3	33.29	48.24	2.22	3.22	6.32	9.17
M4	32.03	46.42	2.14	3.09	6.09	8.82
M5	27.68	40.12	1.85	2.67	5.26	7.62

By observing the above results, it can be seen that the mix containing 5% aluminium dross and 20% of GGBS gives good results and are more proximity to conventional concrete. The increment in the strength is due to proper hydration of cement and reaction of GGBS and aluminium dross. Due to addition of aluminum dross is causes the expansion in the volume of concrete. It is because of the liberation of hydrogen gas which in liberated during the reaction of aluminum dross. Due to the increment in volume of pores in the concrete reduces the strength as well as durability of concrete.

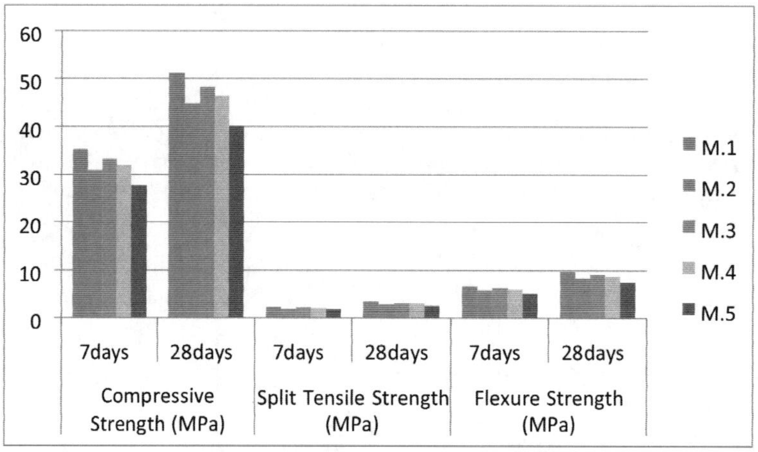

Fig: 1.Mechanical Characteristics

6. Conclusions:

Based on diverse and repeated trials, most advantageous dosage of aluminium dross become decided to be approximately 5%. If cement is changed by using 5 % Al dross in everyday concreting, we can save natural resource (limestone) and additionally reduce pollutants to a degree.

- The compressive power of concrete containing 5% aluminium dross had power in close proximity to that of normal concrete. By the usage of AD in the concrete there is probably discount in the permeability of the concrete.
- The optimized amount of GGBS will affect sand usage inside the construction enterprise. There may be a risk of boom in concrete strength.
- It can be concluded that 20% replacement of GGBS gives maximum bring about strength in comparison to regular concrete after which decreases from 30%.

References

1. Atul Dubey, Dr. R. Chandak, Prof. R.K.Yadav 2012 "Effect of blast furnace slag powder on compressive strength of concrete".
2. M C Nataraja, P G Dileep Kumar, A S Manu and M C Sanjay 2013 "Use Of Granulated Blast Furnace Slag As Fine Aggregate In Cement Mortar".
3. Mohammed Nadeem1, Dr. A. D. Pofale 2012 "Replacement Of Natural Fine Aggregate With Granular Slag - A Waste Industrial By-Product In Cement Mortar Applications As An Alternative Construction Materials"
4. Nesibe Gozde Ozerkan1, Omar Liqaa Maki2, MomenWael Anayeh3, Stian Tangen4, Aboubakr M. Abdullah5 Assistant Professor, 2014.,"The Effect ofAluminium Dross on Mechanical and Corrosion Properties of Concrete".
5. M.Satish reddy, Dr. D. Neeranjan. Dr. B. Kotaiah 2014 "An Experimental Investigation on Use of Secondary Aluminium Dross in Concrete".
6. Gaurav Singha, Souvik Dasa, Abdulaziz AbdullahiAhmeda, Showmen Sahab,SomnathKarmakarb 2015 "Study of Granulated Blast Furnace Slag as Fine Aggregates in Concrete for Sustainable Infrastructure".
7. U. Vamsi mohan, Dr.K.Nagendra Prasad, S. Praveen Kumar Reddy 2015 "Studies on Strength Characteristics of Concrete by Partial Replacement of Sand with Granulated BlastFurnace Slag".
8. Gireesh Mailar , Sujay Raghavendra N , Sreedhara B.M , Manu D.S , Parameshwar Hiremath , Jayakesh K, 2016 "Investigation ofconcrete produced using recycled aluminium dross for hot Weather concreting conditions".
9. Janis Kazjonovs1, Diana Bajare2, Aleksandrs Korjakins "Designing ofhigh density concrete by using steel treatment waste".
10. D Satish Kumar, R Sah, Ganapathi Prasad, S M R Prasad, A Sood, V P Chatterjee and SK Chaturvedi "Granulated Steel Slag - An Alternative ToRiver Sand"
11. Radwa DefallaAbdel Hafez
"The Propertiesand Uses ofAluminum DrossInThe ConcreteIndustry Review".

Travel mode choice changes due to covid-19- case study of Yelahanka

R Akshay[1], Mrs. S Kanimozhee[2] , Mr. Madhu K A[2]
[1]Post Graduate Student, School of Civil Engineering, REVA University, Bangalore-64
[2]Assistant professor, School of civil engineering, REVA University, Bangalore-64
School of Civil Engineering, REVA University, Bangalore-64
(Email: , r.akshay45@gmail.com , kanimozhee.s@reva.edu.in, madhu.ka@reva.edu.in)

Abstract: India had repeated lockdowns from 2020 to 2022, however changes in commuters' travel mode preferences were noted both before and after COVID-19.. In total 950 participant response were created through Microsoft Forms and sent through different forms of social media like WhatsApp, LinkedIn and Facebook and 12% of the responses were recorded out the study area so it was excluded during the study purpose, the total responses were carried out for the study were 827 responses. The study also attempted to quantify the effects of influencing factors which can explain change in the commuter's travel mode choice. The findings revealed that employment growth as seen after COVID-19 of 6% were the main purpose of travelling was to office, recreational and medical services of 19% after COVID-19. Mode choice model were created accordingly for the data analysed. The MLR Model created for Before Covid-19 data shows a good fit and the independent variables are significantly affecting the outcome variable and the Multiple Linear Regression Model created for Post-Covid-19 shows the independent variables are affecting significantly affecting the dependent variable i.e. Mode Choice. Therefore, Multiple Linear Regression Model is finalized and the collected data of the present study shows good model fit. The goal of the current study is to help policymakers better understand the dynamics of shifting travel demand and trip purpose before and after events like COVID-19.

1.*Introduction*

India, the country with the second-largest population in the world with a population of more than 1.34 billion, struggled to stop the spread of the coronavirus 2 that causes severe acute respiratory syndrome among its citizens. Several approaches, such as computer modelling, statistical tools, and quantitative analysis to restrict the spread as well as the quick creation of novel treatments, may be very required to address the current outbreak and its impacts. In order to stop the spread of COVID-19, the Indian Ministry of Health and Family Welfare has made people aware of recent outbreaks and taken the appropriate precautions. In addition, a few lockdowns were ordered nationwide by the Indian government to stop the virus from spreading. The outbreak is also closely tied to the country's economy because it has severely hampered industrial production The COVID-19 pandemic has significantly disrupted people's daily lives, routines, and travel habits all around the world. There has been a lot of continuing research in this field due to the significance of comprehending travel behaviour and mode choice behaviour during COVID-19 times, however the results are still scant. There have been several descriptive data analyses, in particular, on how the various sociodemographic categories of various countries have altered their travel habits in reaction to COVID-19. The relationship between traveller mode switch behaviour and changes in the transport mode utilised to measure the variability associated with the mode switch behaviour. The frequency of travel depends on the various financial parameters of the individuals; those with greater income levels can travel more frequently. Every person's choice of mode of transportation is significant, but employees' commutes depend heavily on their mode of transportation because it makes their travel plans simple and convenient.

2. Methodology

2.1 Methodology:

The methodology of the study starts with framing the objectives required for the study. Then the broad literature survey was carried out related to the study where in the better understanding related to methods and methodology was the first step involved in the methodology is to determine problem identification, due to Covid-19 every individual life style, travel mode preferences has been changed. In order to know theextent to which Covid-19 has affected can find out from this study. After identifying this problem, I have gone through various literature studies conducted in Indiaand around the world. Then studying all the literature papers objectives for this study are defined and study area are being defined. Then the questionnaires are prepared in Microsoft forms and circulated in various social mediaplatforms like WhatsApp, LinkedIn, Facebook, etc. to collect the responses. Due to covid-19 restrictions survey couldn't be conducted physically. After collecting all the responses, the data is going to analyzed through some statistical tests and then models will be developed. Then the models will be validated and concluding the work.

2.2 Flow Chart:

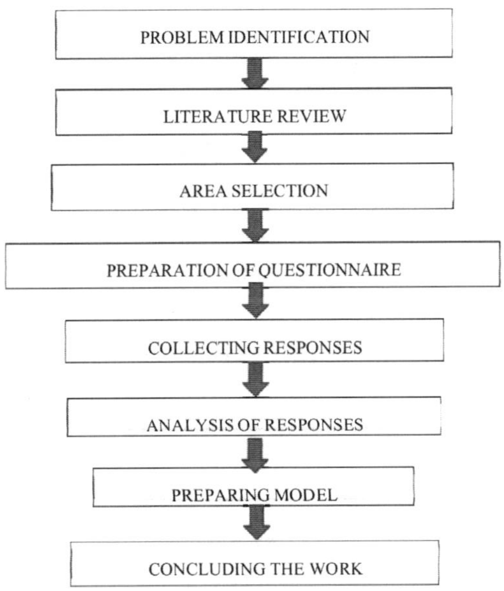

Flow chart 1- Methodology Flow chart

3. Study Area

Yelahanka is a Karnataka suburb of Bangalore, and one of its zones is under the jurisdiction of the BBMP. It is located in the north of the city and is the oldest section of the current Municipal Bengaluru. Yelahanka is traversed by National Highway 44. In addition to passing through the suburbs and the north-eastern edges of Yelahanka, the recently built Navayuga Devanahalli KIAL Highway (Expressway), which is a part of NH 44, also serves as a bypass road to go around the Yelahanka suburb. Yelahanka is located south of State Highway-9, which divides the neighbourhood into Old Town and New Town. There are several attractions in the Yelahanka area. Due to its large undeveloped regions and easy access to the Kempegowda International Airport, Yelahanka is on the verge of becoming the top real estate hotspot in North Bengaluru.

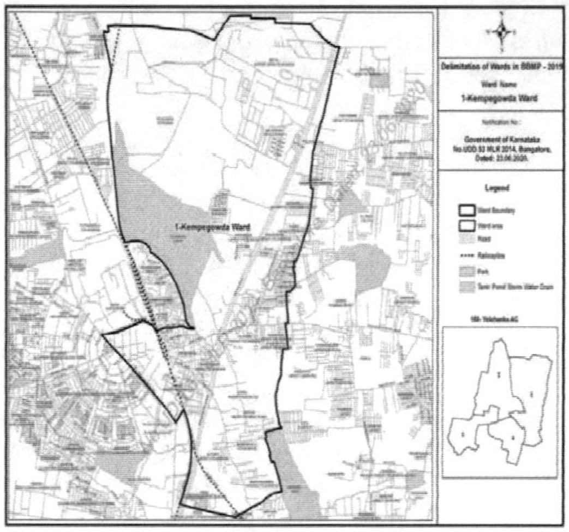

Figure 1. showing study area

Table1: Sample Characteristics of Socio-demographic variable:

Independent Variables: Categorial variables	Sub Categories	Sample Distribution
Age (Years)	Male	36.2
	Female	61.4
	Prefer not to say	2.3
Income (In Rs.)	<25,000	39.3
	26,000-50,000	11.3
	51,000-75,000	36.5
	76,000-1,00,000	9.5
	>1,00,000 And more	3.3
Employment Sector working	Government	7.7
	Semi- Government	23.4
	Private Firms	65.8

3.1 ***Question Survey and Data Collection***

The survey's questions was made using the Microsoft Forms tool. Due to the pandemic's restrictions on face-to-face interviews, the data was gathered online within Yelahanka areas, disseminated via various internet forums including WhatsApp, LinkedIn, and Facebook. A couple of my friends helped me spread the questionnaire link on social media platforms. Replies were gathered using a random sample technique, and a total of about 827 responses received. The major reason individuals used to take their trips was described as the principal purpose of travel. During a pandemic, people seem to be able to cut back on a few other crucial travels, but they could also still travel for a specific major trip objective. It is crucial to pay close attention to the main reason for travelling because it affects the frequency of journeys, the distance covered, and the method of transportation used. Prior to and following the Covid-19 pandemic, questions about the main reason for travelling were included in a section of the questionnaire. Respondents' levels were determined using a 5-point rating scale, and their comfort levels for using public transportation before and after the Covid-19 survey were determined using Net Promoter Score analysis. These factors included travel time savings, economy, safety and security, comfort, and social distance.

4. Discussions:

Figure 2. Travel Frequency used Work from home Before Covid-19

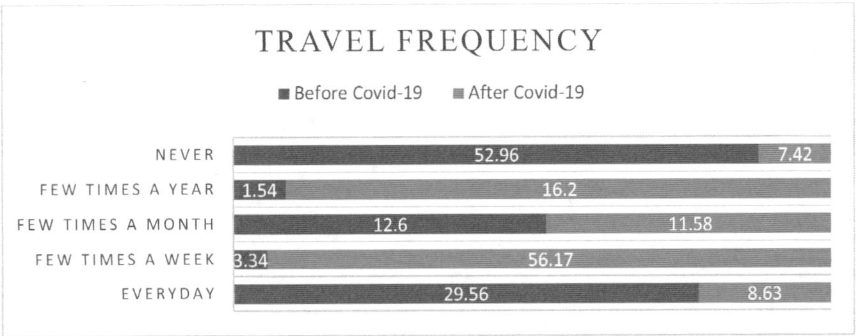

Impact of covid-19 has changed the work structure for many people in the above figure 2 we can see that many people have not faced work from home scenario before, maximum of (52.96%) said never had been in such a situation. Many people had been to the office every day before covid-19 of 29.56% and the rest combined 17.48% visited a few times a week, month and year.
Impact of covid-19 has changed the work structure for many people in the above figure 5 we can see that many people have faced work from home scenario after Covid-19, maximum of (59.21%) said that they had been too few times a week after Covid-19 , few people had been to the office few times an year after Covid-19 of 20.72% and the rest combined 20.07% visitedevery day, few times a month and never been to office after Covid-19.

4.1 Multi Linear Regression Model

4.1.1 Multi Linear Regression Before Covid-19

Table 2: Pearson's Correlation Analysis

Variables	r	P
Travelling mode before Covid-19	1.000	.
Age	0.39	.283
Gender	0.84	.049
Highest Degree	-.129	.005
Income	-.222	.000
Two-wheelers	.32	.046

According to the objective of the study the outcome of the variable 'Travel Mode choice Before Covid-19 must be determined, hence it is selected as Independent Variable in Multiple Regression modelling. Table 1 exhibits the co-efficient of rest others variables with 'Travel Mode choice Before Covid-19' and their level of significance. The Independent variables must be correlated with the outcome variable with co-efficient >0.3. Therefore, the variables which are significant and suitable to be Independent

Variables are: Age, gender, highest degree of qualification, income, monthly expenditure spent, travelling frequency before covid-19. Hence 16 Independent Variables have better correlation with the Dependent variable. To check the next assumption of collinearity i.e. Absence of Multi-collinearity among the Independent variables, the correlation co-efficients among themselves must be checked. The correlation co-efficients of gender, Income, travelling frequency were found out to be significant of .049,.000, and .000 respectively. If the co-efficient values are >0.7 among any pair of variables then the Multi-collinearity among those pair of Independent variables exists can be excluded for the analysis purpose.

The anova shows that dependent and the independent variables are having a good relation and most of the independent gave given good model output so as the significancevalue is (.001) which is <.05 we can say the fit is a good choice. The Model summary shows the **R square value** is interpreted from the model summary it explains the variance of the Independent variable with respect to dependent variable and the value of point(.009) is obtained for the present study which is considered fairly good and also has better significance value which is lesser than .05.

The Standard residual in the Residual statistics must not beyond the values -3.3 to +3.3. The Standard residual value obtained in the statics is -2.455 and 1.522 for the current data indicating no outliers. The maximum Cook's distance is also not greater than 1 which another indicator of absence of outliers.

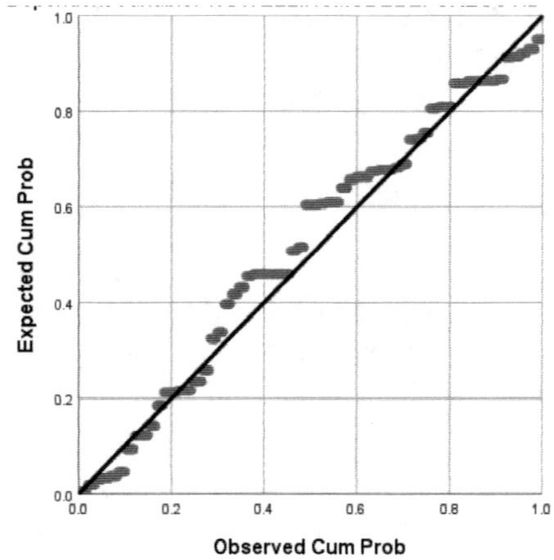

Figure 3: Normal P-P Plot of Regression Standardized Residual

4.1.2 Multi Linear Regression After Covid-19

Table 3: Pearson's Correlation Analysis

Variables	r	P
Travelling mode before Covid-19	1.000	.
Gender	-.151	.001
Highest Degree	-.014	.001
Employment work structure after Covid-19	.42	.043
Frequency of working from home afterCovid-19	.35	.004
Travelling frequency to work after Covid-19	-.42	.000

The Independent variables must be correlated with the outcome variable with co-efficient >0.3. Therefore, the variables which are significant and suitable to be Independent Variables are: Age, gender, highest degree of qualification, employment work structure after Covid-19, monthly expenditure spent, travelling frequency After covid-19. Hence 16 Independent Variables have better correlation with the Dependent variable.
To check the next assumption of collinearity i.e. Absence of Multi-collinearity among the Independent variables, the correlation co-efficient among themselves must be checked. The correlation co-efficient of gender, highest degree, travelling frequency were found out to be significant of .001,.001, and .000 respectively.The anova shows that the dependent and the independent variables are having a good relation and most of the independent gave given good model output so as the significancevalue is (.001) which is <.05 we can say the fit is a good choice.
The Model summary shows that **R square value** is interpreted from the model summary it explains the variance of the Independent variable with respect to dependent variable and the value of point(.048) is obtained for the present study which is considered fairly good and also has better significance value which is lesser than .05 The Standard residual in the Residual statistics must not beyond the values -3.3 to +3.3. The Standard residual value obtained in the statics is -2.030 and 2.150 for the current data indicating no outliers. The maximum Cook's distance is also not greater than 1 which another indicator of absence of outliers.

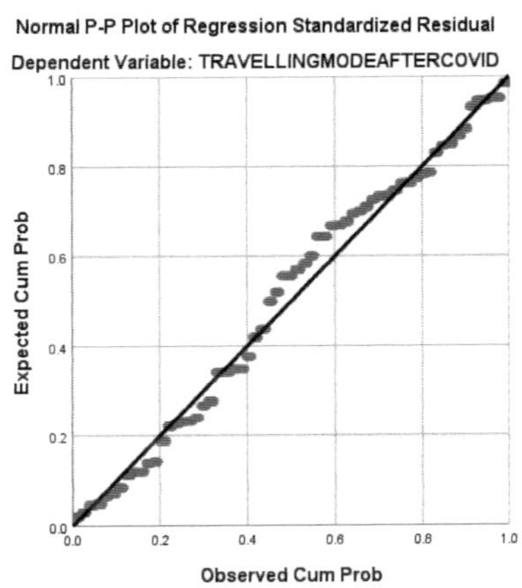

Figure 4: Normal P-P Plot of Regression Standardized Residual

The above plot of P-P graph (figure 4) depicts the Linear distribution of the considered variables and it is seen that the points nearly following the line of diagonal although there are little deviations, which is the identification of linear distribution of variables.

The Equation of Multi Linear Regression is as follows;
$Y_{modechoice}=0.219X_{AG}+0.113X_{GE}-0.089X_{IN}+0.005X_{HD}-0.017X_{EMP}+0.040X_{EMW}+0.110X_{TF}+0.329X_{PT}-0.186X_{MTC}+0.084X_{PTT}$

Where, $Y_{mode\ choice}$ is Dependent Variable of the Multiple Regression Equation
X_{AG} is Independent Variable of Age
X_{GE} is Independent variable of gender
X_{IN} is Income
X_{HD} is Independent variable of Highest degree
X_{EMP} is Independent variable of Employment status before/after Covid-19
X_{EMW} is Independent variable of Employment work structure before/after Covid-19
X_{TF} is Independent variable of Travel frequency to work before/after Covid-19
X_{PT} is Independent variable of Usage of public transportation before/after Covid-19

X_{MTC} is Independent variable of Monthly travelling cost before/after Covid-19
X_{PTT} is Independent Variable of Primary travel after before/Covid-19

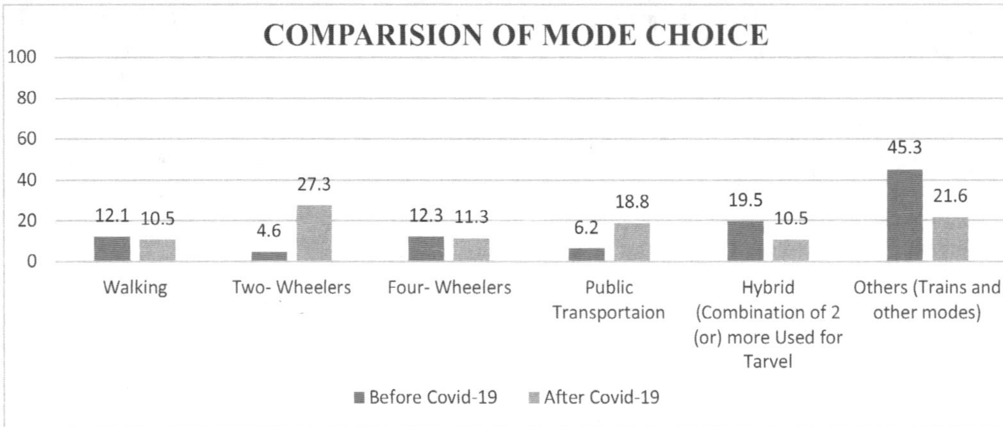

Figure 5. Comparisons of Mode choice before and After Covid-19

5. Conclusion:

From the data, it is observed that employees in Yelahanka town has been affected to what extent Covid-19 affected the preference of Mode choice for various trip purposes of travel MODE Choice has affected on the employees in Yelahanka Town. From the above graph we can see that there is changes in the mode choice of employees in the Yelahanka town before and post Covid-19. We can see that there is increase in the usage of two-wheelers of (27.3%)and usage of public transportation has seen increased post Covid-19 (18.8%).The Multiple Linear Regression Model created for Before Covid-19 data shows a good fit and the independent variables are significantly affecting the outcome variable and the Multiple Linear Regression Model created for Post-Covid-19 shows the independent variables are having good significance affecting the dependent variable i.e. Mode Choice. Therefore, Multiple Linear Regression Model is finalized to interpret the collected data of the present study as it shows good model fit in compression with Multinomial Regression model.

6. References:

1. M. Abdullah, C. Dias, D. Muley, M. Shahin, *Exploring the impacts of COVID-19 on travel behavior and mode preferences*, Transp. Res. Interdiscip. Perspect. **8** (2020) 100255.
2. M. Abdullah, A.L.I. Nazam, S.A. Hussain, A.B. Aslam, M.A Javid, *Measuring changes in travel behavior pattern due to COVID-19 in a developing country: a case study of Pakistan*, Transp. Policy (Oxf) **108** (2021) 21–33.
3. M. Abdullah, N. Ali, C. Dias, T. Campisi, M.A. Javid, *Exploring the Traveler's inten- tions to use public transport during the COVID-19 pandemic while complying with precautionary measures*, Appl. Sci. 11 **(8)** (2021) 3630.
4. M. Abdullah, M.A. Javid, *Accessibility-based approach: shaping travel needs in pandemic situation for Planners'* perspectives, Eng. J. 25 **(1)** (2021) 15–22.
5. A. Aloi, B. Alonso, J. Benavente, R. Cordera, E. Echániz, F. González, R. Sañudo , *Effects of the COVID-19 lockdown on urban mobility: empirical evidence from the city of Santander (Spain)*, Sustainability 12 **(9)** (2020) 3870 .
6. M.A. Al-Rashid, H.C. Goh, Y.A.S. Harumain, Z. Ali, T. Campisi , T Mahmood , *Psy- chosocial barriers of public transport use and social exclusion among older adults: empirical evidence from Lahore,*

Pakistan, Int. J. Environ. Res. Public Health 18 **(1)** (2021) 185.

7. D.M. Barbieri, B. Lou, M. Passavant, C. Hui, I. Hoff, D.A. Lessa , T.H. Rashidi , *Impact of COVID-19 pandemic on mobility in ten countries and associated perceived risk for all transport modes*, PLoS ONE 16 **(2)** (2021) e0245886 .
8. R. Basu, J. Ferreira, *Sustainable mobility in auto-dominated Metro Boston: chal- lenges and opportunities post-COVID-19*, Transp. Policy (Oxf) **103** (2021) 197–210.
9. P. Bucsky, *Modal share changes due to COVID-19: the case of Budapest*, Transp. Res. Interdiscip. Perspect. **8** (2020) 100141.
10. R. Buehler, J. Pucher, *Demand for public transport in Germany and the USA: an analysis of rider characteristics*, Transp. Rev. 32 **(5)** (2012) 541–567.
11. M.A. Bujang, N. Sa'at, T.M.I.T.A Bakar, *Sample size guidelines for logistic regression from observational studies with large population: emphasis on the accuracy between statistics and parameters based on real life clinical data*, Malaysian J. Med. Sci.: MJMS 25 **(4)** (2018) 122.
12. T. Campisi , S. Basbas , A. Skoufas , N. Akgün , D. Ticali , G. Tesoriere , *The impact of COVID-19 pandemic on the resilience of sustainable mobility in Sicily*, Sustainability **12** (21) (2020) 8829 .
13. A. Carteni, L. Di Francesco, M. Martino, *how mobility habits influenced the spread of the COVID-19 pandemic: results from the Italian case study*, Sci. Total Environ. **741** (2020) 140489.
14. D.K. Chu , E.A. Akl , S. Duda , K. Solo , S. Yaacoub , H.J. Schünemann , M. Reinap , *Physical distancing, face masks, and eye protection to prevent person-to-person transmission of SARS-CoV-2 and COVID-19: a systematic review and meta-analy- sis*, The Lancet 395 **(10242)** (2020) 1973–1987 .
15. T.S. Combs, C.F. Pardo, *Shifting streets COVID-19 mobility data: findings from a global dataset and a research agenda for transport planning and policy*, Transp. Res. Interdiscip. Perspect. 9 (2021) 100322.
16. S. Cullinane, *The relationship between car ownership and public transport provision: a case study of Hong Kong*, Transp. Policy (Oxf) 9 **(1)** (2002) 29–39.
17. M. de Haas, R. Faber, M. Hamersma, *How COVID-19 and the Dutch 'intelligent lock- down'change activities, work and travel behaviour: evidence from longitudinal data in the Netherlands*, Transp. Res. Interdiscip. Perspect. 6 (2020) 100150.
18. D. Deka, *Transit availability and automobile ownership: some policy implications*, J. Plann. Educ. Res. 21 **(3)** (2002) 285–300.

A Study on Natural Clay and Moringa Oleifera for Sustainability of Landfill Liners

Pavithra M P[1], Dr. R L Ramesh[2], Sandya Rani [3]
[1]Assistant Professor, School of Civil Engineering, REVA University, Bengaluru – 560064, Karnataka, India.
[2]Professor, Department of Civil Engineering, Don Bosco Institute of Technology,Bengaluru-560074, Karnataka, India
[3]Assistant Professor, Department of Civil Engineering, Don Bosco Institute of Technology,Bengaluru-560074, Karnataka, India.. (pavithra.mp@rea.edu.in ; rameshrantaval@gmail.com ; sandhyaranigms@gmail.com)

Abstract. This project aims to characterize the municipal solid waste leachate, using column study method to evaluate the concentrations of constituents present in landfill leachate. The leachate sample were tested for several physical and chemical parameters. The initial results demonstrated that the characterizations of leachate found to be highest range of contamination. Hence compacted clay liner is used in this treatment. Moringa Oleifera is added with compacted natural clay to retain the constituents in the liner. Compacted natural clay liner shows the removal efficiency of TDS, Sodium and Potassium is 88.44%, 90.6% and 68.46% respectively. Moringa Oleifera with Compacted natural clay liner shows the removal efficiency of TDS, Sodium and Potassium is 90.2%, 98.75% and 81.2% respectively. Result shows Moringa Oleifera with Compacted clay liner has been an effective for the removal of constitutes present in landfill leachate.

1. Introduction

The frequency of municipal solid waste production in large cities is increasing as a result of rising population and rapid economic growth. Local governments are trying to establish a proper solid waste management system that meets their needs. MSWM (municipal solid waste management) has emerged as a major environmental concern, particularly in developing countries. Bengaluru, Karnataka's state capital, is one of Asia's fastest developing cities. With a land area of 2190 km^2 and a population of roughly 10.18 million, the Bruhat Bengaluru Mahanagara Palike (BBMP) creates around 5000 metric tons of solid garbage every day at a rate of 0.5 kg per capita per day (kg/capita/d). Bengaluru City is currently experiencing significant problems as a result of existing waste disposal practices, which are costing the city a lot of money due to a lack of proper infrastructure; additionally, the structures built on these former dumps face major issues as a result of open dumping in the city's expanding zone. Meanwhile, due to poor leachate treatment, groundwater quality has deteriorated. Intention is to examine potential environmental implications and recommend to improve the trash management methods.

The present paper discusses, the feasibility of natural clay and Moringa Oleifera used as landfill liners were evaluated. Their physiochemical characteristics were addressed. Finally, the retention capacity in leachate was evaluated for sustainable waste landfill applications. The main goal was to create a landfill liner that was cost-efficient while yet being geotechnical stable and effective at attenuating pollution. Leachate spilling over nearby rivers, lands, or groundwater aquifers could cause environmental contamination and harm to human health.

- To study the environmental sustainability of landfill.
- To check the feasibility of natural clay available at the landfill.
- To find out the physio-chemical characteristics of leachate and properties of soil.
- To increase the retention capacity of the natural clay liner by adding Moringa Oleifera.

2. Materials and Methodology

2.1 Materials:

Soil is obtained from Bellahalli dumpsite in Bangalore was used in this study. The samples were collected by open excavation from a depth of 1 meter from natural ground. The total sample of 25kgs is collected from the site. The soil so obtained has clay and silt content of 62%. The properties of soil is tested.

Figure 1. Natural clay soil

Leachate is collected from the solid waste landfill located at Belahalli for this study. The sample is taken from freshly collected leachate in site. The primary causes of varying concentrations in various landfills are changes in temperature, the state of the climate, the soil used for land filling, the organic and inorganic components found in solid waste, the waste's age, other features, and the manner of land filling.

Figure 2.Leachate liquid

Removing the coat and wings of Moringa Oleifera, seeds were mechanically dried and dehusked. The white seed kernels were ground up and put through a sieve with a mesh size of 0.75. To obtain the natural coagulant extracted from moringa oleifera, the seed coats were removed and were blender crushed until obtaining a powder form, the powder was weighed for 300gms and was mixed with the natural clay for further leachate collection by concentration. From the neighbourhood market in Bengaluru, India, moringa oleifera seeds were bought.

Figure 3. Moringa Oleifera Powder

2.2 Methodology:

The investigation used a vertical, cylindrical column that was 4 cm in diameter and 11 cm deep. The column's flat bottom and maintained opening are at its top. At the outlet, screens are positioned to collect effluent free of any solid particles. A steel platform supported the column, which was filled with soil samples. In order to maintain flow rate and prevent clogging, a fine screen is installed at the exit of the column on top to distribute landfill leachate onto the surface of the adsorbent. The column was known as a bed of compacted earth column. The test setup was maintained for a total of 28 days. Collected leachate is measured for every day. For each 50 ml sample, an effluent test was conducted continuously to check out the parameters such as Sodium, Potassium and total suspended solids.

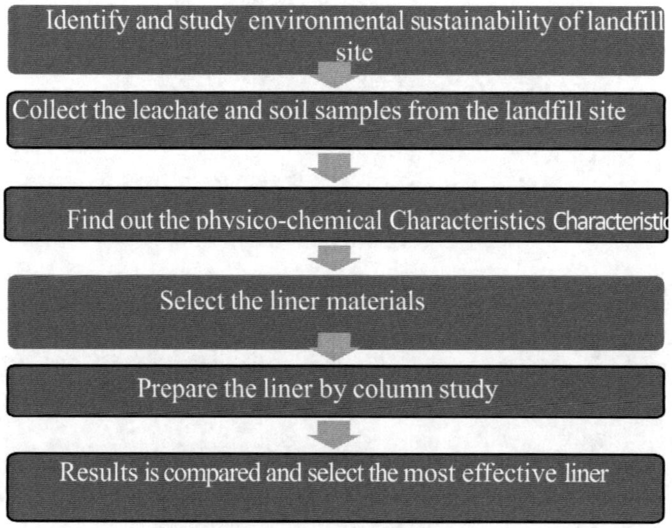

Figure 4. Flow chart

The results from the treatment of the leachate from the column research are examined for their physical-chemical traits like sodium, potassium, and total dissolved solids (TDS) for the 11 cm depth of the soil-packed column.

3. Result and Discussion

Table 1: Chemical characteristics of Leachate sample

Sl no.	Parameters	Unit	Method	Results	Standards values
1	Total chromium as Cr	mg/L	ICP-OES	0.153	2.0
2	Copper as Cu	mg/L	ICP-OES	3.176	3.0
3	Zinc as Zn	mg/L	ICP-OES	1.392	15
4	Sodium as Na	mg/L	IS 3025 (Part-45)1993	1762	200
5	Potassium as K	mg/L	IS 3025 (Part-45)1993	1897	10-30
6	Lead as Pb	mg/L	ICP-OES	0.174	1.0

Table 2: Physical characteristics of Leachate sample

Sl No.	Parameters	Unit	Method	Result	Standard values
1.	PH		IS 3025 (Part-11):1983	7.95	5.5-9.0
2.	Total Dissolved solids	mg/L	IS 3025 (Part-14):1984	13075	250-850
3.	Total Hardness as CaCO3	mg/L	IS 3025 (Part-21):2009	2746.46	600
4.	Fluoride as F	mg/L	APHA23rd Edition4500-SO4:2017	1364.67	15
5.	Chloride as Cl	mg/L	IS 3025 (Part-32):1988	3176.32	1000
6.	Chemical Oxygen Demand as O2	mg/L	IS 3025 (Part-58):2006	1680	250
7.	Nitrate as NO3	mg/L	APHA23rd Edition4500-SO4:2017	43.45	45
8.	Sulphate as SO4	mg/L	APHA23rd Edition4500-SO4:2017	105.55	1000

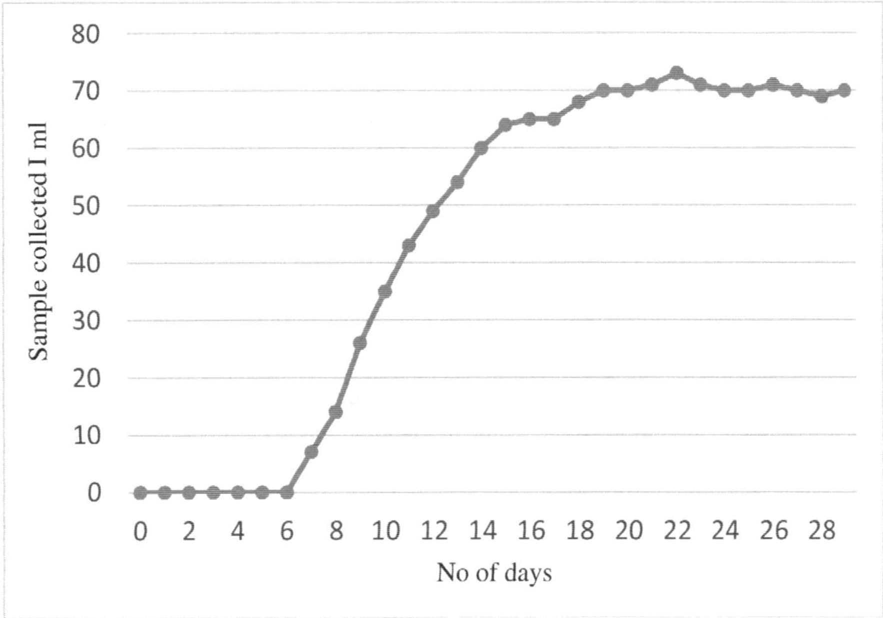

Figure 5. Sample collected through compacted natural clay liner for 0-29days

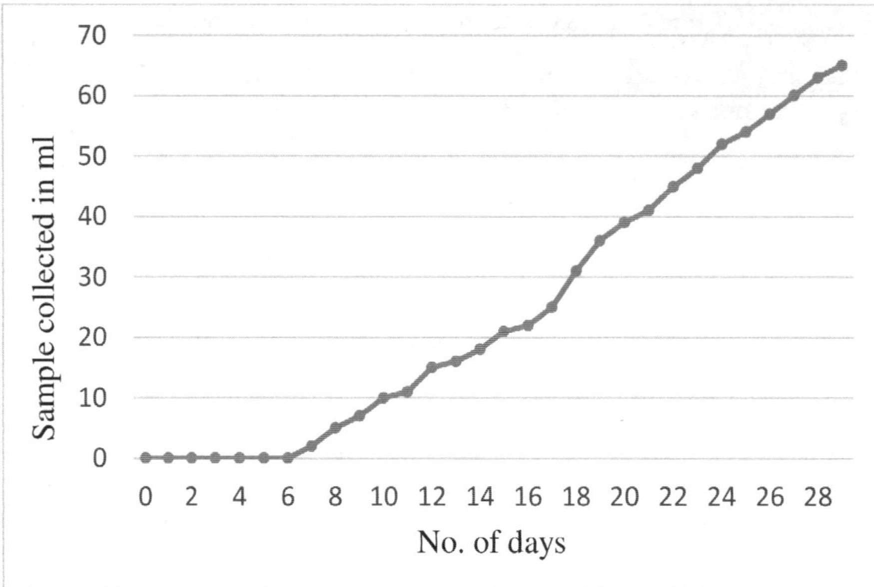

Figure 6. Sample collected through Moringa Oleifera liner for 0-29days

Table 3 Test results of collected leachate liquid of Natural compacted clay

Sl. No	Parameters	Initial value	Final value (mg/l)	Removal Efficiency in %
1	Sodium	1762	203.6	88.44
2	Potassium	1891	178.3	90.6
3	TDS	13075	4123.4	68.46

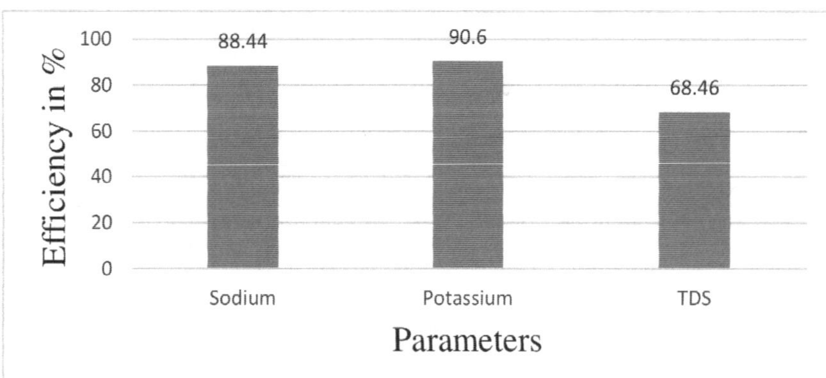

Figure 7. Efficiency of different parameters of compacted clay liner

Table 4 Test results of collected leachate liquid of Moringa oleifera

Sl. No	Parameters	Initial value	Final value (mg/l)	Efficiency in %
1	Sodium	1762	171	90.2
2	Potassium	1891	23.5	98.75
3	TDS	13075	2452.5	81.2

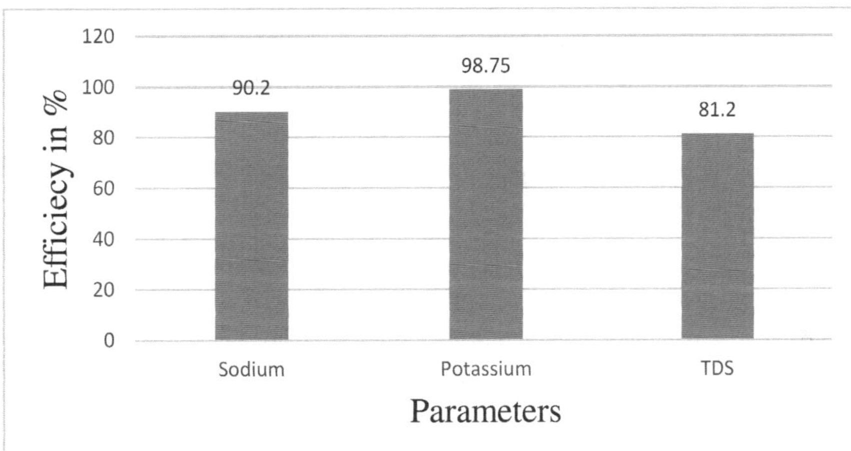

Figure 8. Efficiency of different parameters of moringa oleifera

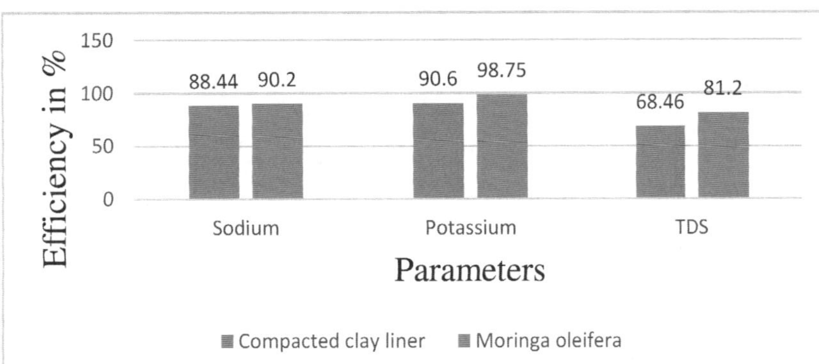

Figure 9. Graphical representation of natural clay and moringa oleifera comparison

This graphical representation shows the comparison of natural clay and moringa oleifera therefore, the results conclude that moringa oleifera has a high efficiency compared to natural clay.

4. Conclusion:

From the present study the following general conclusions can be drawn

• Its properties were examined when a sample of leachate was collected. A significant amount of sodium, potassium, and total dissolved solids, among other substances, are thought to be present in the leachate sample.

• Moringa oleifera and compacted clay liner is selected to improve the feasibility and retention capacity of the natural clay liner. Moringa oleifera showed better result than compacted clay liner.

• There is an advantage by using economically available moringa oleifera as it can reduce the cost and is environmentally friendly. The result of present analysis shows that, this treatment using moringa oleifera has been an effective for the removal of constitutes from the leachate.

5. References

[1] Amin Mojiri, John L. Zhou, Harsha Ratnaweera, Akiyoshi Ohashi, Noriatsu Ozaki, Tomonori Kindaichi and Hiroshi Asakura (2020) – *Journal of Water Reuse and Desalination* – Treatment of landfill leachate with different techniques

[2] Witold Stępniewski1, Marcin K. Widomski1 and Rainer Horn – Lublin University of Technology, Faculty of Environmental Engineering, Dept of Land Surface Protection, Christian Albrechts University, Institute for Plant Nutrition and Soil Science, Poland (2011) – *Hydraulic Conductivity and Landfill Construction*

[3] Agus Muntohar (2001) – Universitas Muhammadiyah Yogyakarta – A comparative study of different additive on the index properties of expansive soils

[4] Vasyl Popovych1, Jerzy Telak*, Oksana Telak, Myroslav Malovanyy, Roman Yakovchuk1, Nataliya Popovych (2020) – *Migration of Hazardous Components of Municipal Landfill Leachates into the Environment.* Volume21, Issue1, pages 52-62

[5] Ayoola Nike Daniel, Ike Kenneth Ekeleme, Chinemerem Martin Onuigbo, Victor Okezie Ikpeazu and Smart Obumneme Obiekezie Department of Microbiology Nasarawa State University Keffi, Nigeria (2021) – *Review on effect of dumpsite leachate to the environmental and public health implication*

[6] Syed Abu Sayeed Mohammed; Maya Naik; Sanaulla Pathapalya Fakhruddin Zulfiqar Ahmed Mohammed Nazeer (2008) – *An Interdisciplinary Journal of Applied Science* – Studies on contaminant transport at an industrial waste dumpsite of Bangalore, India

APPLICATION OF BIM FOR EFFECTIVE CONSTRUCTION SAFETY MANAGEMENT IN HIGH RISE BUILDINGS

Swaroop Devaiah K. L.[1]**, Vasanth Keshav**[1]

[1]**School of Civil Engineering, REVA University, Bangalore**
Corresponding Author: keshav.vasanth@gmail.com

Abstract: The construction industry has a higher occupational casualty rate than other industries. As a proactive approach to safety management, a good safety management system (SMS) can go a long way to help prevent accidents and occupational hazards. This work proposes the use of BIM 3D with the help of Autodesk Revit for the identification of hazards and risks at the designing stages and for implication and planning of effective safety measures. This is carried out from the inception of the project itself which later can be applied to the construction process based on efficient SMS. The SMS established is based on surveys conducted on 45 personnel in the construction industry. The top 5 important issues found to be associated with site safety were: Exposed Rebars; Laboures and Visitors without ID; CCTV monitoring of site; Signs not secured properly; Access to Fire Engine and personnel in case of fire break out.

Keywords: Building Information Modelling [BIM-3D], Construction Safety, High Raise Structure, RRI.

1. Introduction

Throughout history from the bigging one of the riskiest industries to worn in has been Construction Industry. There are various risks which occur throughout the industry it could be a fall down, a machine malfunction or mis control, being exposed to harsh environment, etc. A good quality [SMS] Safety Management System can be a dementing factor to help avert accidents and industrial hazards. A useful SMS goes further than the idea of, "do this, don't do that" attitude towards safe construction practices. In the time construction managers were still utilizing notepads and paper and slider ruler, and prior to construction management software made its first appearance within the sector, initial safety management systems were in use already. During the safety regulations during the period, it was "one size fits all."
The assumption for the safety of construction was based whether the company was legal or not. As time changed, the way of safety management & its strategy has also changed. Drastically, many construction enterprises have considered within the sector are three main important reasons to need an operational safety management system:

- Moral obligation. No corporation should place its members, or people associated with the work of the corporation, at risk.
- Regulations. Many states and countries make it a legal obligation for companies to encourage and maintain secure working environments.

- Cost-effectiveness. Time, energy, and money spent on averting accidents is less than the expenditure of dealing with accidents later.

In Contrast, what safety management does, a safety management system for construction industry is a methodical way of detecting risks and managing risks relating to the construction work. The Safety Management System involves organizational operation, responsibilities, systems, procedures and construction corporation's policies for assuring that the essential safety measures are being applied and are being retained for the safety of all involved. Further above that, the SMS needs to be rooted in the culture of the enterprise, to make sure its been followed by all.

[1] investigated the factors affecting safety on construction sites in terms of historical, economical, psychological, technical & organisational and to give suggestions to improve safety scenario at the construction sites. The methodology adopted was to interview various stakeholders through questionnaires. They concluded that variables related to 'organisational policy' are the most prevalent group of factors affecting safety performance. [2] studied the usefulness of the safety management events as at present adopted by Australian governments. The paper details on a very detailed empirical analysis conducted to examine the correlation between the amount of safety management responsibility and the overall safety management. The methodology adopted was to survey on med & large construction organisations. The author concluded that less Pro-activeness is there on construction sites to improve safety. It also states that only 25% felt that policy of appointing safety personal is excellent. [3], carried out extensive research work to distinguish where safety is bargained and why. They established a causality model, with a combination of focus groups and comprehensive study of 100 construction accidents, utilizing an ergonomics systems method. They concluded that the stakeholders are responsible for the direct accident circumstances.

[4], investigated the safety status in building construction sites in Kolkata. The methodology adopted was to survey the construction sites. They conclude that despite existence of Indian safety codes, the sites were found to be deficient in employing the safety requirements. Also, it states that the economy of the company and the country also plays a significant role in implementing safety in construction sites. [5] studied the current scenario of Indian labours working in construction industries in terms of health & safety issues. The methodology adopted was to survey the construction sites. In conclusion he stresses on the widened gap between the ground realities and the safety principles. In this regard, he stresses on better implementation of existing government laws. [6] studied and developed and safety checking platform based on specified rules which involved edge detection in BIM. They also succeeded applying this program on two case studies. Through this they were able to identify potential hazards involving risk of fall from the inception of the project and adjustments were made based the visualization results.

[7] focused on the comprehensive study on "Safety management in the construction industry, identifying the risk and reducing accidents to improve safety productivity and ROI of the company". The study revealed that crucial industry movements, such as the use of BIM and pre-fabrication are having potent positive impacts onto project safety. [8] emphasized on the importance of construction safety and talks about various progress in technology regarding the same like automatic rule checking in BIM. He also talks about the lack of adoption of technologies are low due to the low number of experts in the field and low trained and skilled employees in the same, low requirement from clients regarding safety and cost of software. [9] discussed various ways for construction safety through BIM. The authors concluded that most of the designers don't have a method for identifying such risks. In this study they developed an DFS based library to provide a knowledge guide for designers. Through the process it was evident that such a tool was useful in automatic identification of risks in the design.

Reference [10] did a detailed study on various literature dedicated to BIM. For this reason and detailed scient-o-metric and bibliometric study conducted on 97 articles. A relation between BIM features and Safety factor was represented through an feature-factor matrix. It was concluded that there is a lack of research into BIM in safety management.

[11] introduced a system called indoor positioning system (IPS) and inertial measurement unit (IMU) which works along with BIM to provide various details of the workers who have equipped the necessary instrument. This provided detail such as their location, height they are working in moving speed direction they are facing, etc. [12] suggested a unique technique of quantitative construction safety risk assessment for construction projects at the design stage. This is done by considering three indicators: Exposure, consequence and likelihood. These indicators are evaluated using construction site data which is objective and accurate, work-related deaths and injuries. In Autodesk Revit a link between safety risk information and BIM is created using an plugin, which is capable of helping structural designers and architects to select alternative designs by calculating the construction safety. [13] conducted an bibliometric study into the construction safety management technologies. In this study they looked into various technologies like robotics, 4D CAD, UAV, database, GIS, VR, BIM. A total 240 papers pertaining to the topic were studied. [14] through the study wanted to improve the Health and Safety in construction site. For this purpose, they interviewed and surveyed various people working in the industry. They understood that BIM is an very suitable tool for increasing safety but as multiple obstacles to overcome. They concluded although a lot of work is done there is no significant software available for this purpose. Few countries have been able to develop techniques for this task but is not widely available. [15] conducted a bibliometric study in economics of construction safety. The authors concluded that developing countries have more HS papers than developed countries suggesting more construction events and problems in such countries. Also, how deep learning and AI could play very important role in future of HS but would require significant financing are highlighted by the authors.

In this study the SMS (Safety Management System) for high-rise construction structures utilizing B.I.M. Visualization and Revit architecture is established. Further planning of Jobsite Work plan Layout using Building Information Modelling – B.I.M. tools for better safety management systems – SMS is conducted.

2. Methodology

The methodology adopted for the realisation of the project on "crucial factors identifications and Implementation of BIM" has been explained in the following flow chart. First, we conducted Literature Review, on the basis of which a questionnaire was prepared and stakeholders from construction industry were interviewed. On the basis of these interviews crucial factors affecting the safety at construction site were identified. Based on the lacuna a Jobsite Workplan Layout was made and recommendations were given.

Different safety manuals which are being used as a safety checklist for the construction site, previous report on safety management in construction have been referred to prepare the questions for surveys. The questions were categorized in their respective fields. Different levels of management like Top level management, Project managers, Senior Engineers, Site/assistant Engineer and Safety Engineers were to be interviewed. The fields which have been considered for preparing questions have been mentioned in Table 1:

Table 1 Shortforms of Different Fields Used For Analysis

Jobsite general	JG
Electrical	El
Excavation/Earthwork	Ex/Ea
Safety netting	SN
Fire risk & management	FR
Fire prevention	FP
Flammable liquid/ materials	FL
First Aid	FA
Housekeeping & sanitation	HKS
Lifting operations	LO
Hoists, Crane & Derricks	HC
Personal protective equipment	PPE
Road works	RW
Construction areas- secured access/after hours	CA
Scaffoldings/Working platform	SW
Site set-up	SS
Traffic management	TM
Working at heights	WH
Ladders	L
Welding & cuttings	WC
Concrete construction	CC

Data collection is done through Microsoft forms. The collected data will be analysed using Relative ranking index (RII), in Statistical Package for Social sciences (SPSS) software. RII gives important factors based on ranking and further analysis is done by understand where safety is prioritized and where its being ignored.

3. Results and Discussions

3.1. RRI method

Relative rank index is utilized for contrasting between the significances of each factor. The factors are addressed with Likert scales which range from 1 to 5 clarifying the significance of every factor chosen by respondents and should be converted into a RRI that has a value of one or less. A Four-point Likert scale or Five-point and Seven-point Likert scale can be utilized for examination for the most part. A Five Point Likert Scale is used for the analysis. From the RRI method the following points are noted.

- It was thus derived from the responses and through analysis that the most neglected safety precaution taking place in construction industry is open ended rebars i.e. rebar caps are not being utilized on sites.
- The second most significant factor is ID badges not being issued foe labourers and visitors to site. This is a significant disregard to safety because if a pedestrian unknowingly enters and construction site and is not aware of the site safety guidelines may come into danger. Also not recoding and movement of people entering and exiting site may lead to cases of theft and any

accidents or unfortunate incidents taking place may fall under the responsibility of Contractor or Client

- Improper or no CCTV monitoring on construction site is again another major reason which is contributing to inadequate safety monitoring on construction site. It is very important to have CCTV monitoring on every construction site regardless of the scale of project.
- Although safety signs and road signs are placed in the construction site, in most cases they are not secured properly. This is especially important for road signs as they help avoiding any traffic related incidents.
- Segregation of waste is another important part of construction safety as well as for environment. Through the survey its understood that combustible and non-combustible waste are being disposed into same bin.
- All construction site needs to have a fire engine access road, although there are roadways which allow access for heavy vehicles these road-way do not access all parts of a construction site. Proper precaution is needs to make sure a fire access road is made available as freely as possible.
- It was found that although safety signs and road signs are put up, no smoking signs are insufficient in numbers. This is very important as many flammable liquids and materials are utilized throughout the construction industry.
- In many construction sites it was discovered through the survey that although hazardous waste is being generated no proper policy or procedure is put in place for handling of such waste.
- Although most construction sites have fire risk assessment and have procedure for evacuation in case of fire break out, from the survey it is noted that not all staff and labourers are trained for it.
- Within construction site ladders are used for multiple jobs within the construction activities. Although ladders are in use they are not properly inspected, this is a major problem for construction safety as falling from heights is one of the most common accidents taking place in construction site.

3.2. BIM for Construction Safety Planning Work

BIM permits assessment of project approach and sequencing, and therefore recognize the potential safety dangers across the construction ventures. Let's recognize how BIM and 3D-Based Integrated Solution help safety construction for contractors. BIM accepts an efficient visual interpretation of the working conditions for construction, even before the first brick is laid. Construction site staff and new workers can appreciate the safety measures put in place at the construction area and also understand how to avoid circumstances that puts their safety at stake, by using BIM visualization. Also, BIM helps realize the construction practices faster and better while safety concerns are being addressed. As discussed previously, there can be numerous potential safety hazards for a construction project. Recognizing these possible hazards, evaluating the "what if" circumstances and the risks concerned, safety plan choreography with the help of mangers and efficiently handle "what if" situations, as and when they happen without triggering any potential damage to property or life.

Currently, the basic aim of identifying and assessing potential risks and construction site conditions is to sketch our safety implementation program. Nevertheless, BIM provides more benefit than this. Moreover, it gives an overall understanding of the construction method, and permit engineers, view every schedule and the complicated processes in great order, and hence prepare for each activity in a better means. The points that needs to be deemed important for construction safety planning include:

- The project planned and its location
- The placement of machinery which are heavy such as BBR machine and cranes
- Planning for the excavation process
- Planning and designating location for storage of pre-fabricated elements, construction components, unloading and loading
- Disposal and collection of hazardous waste matter produced during construction

Table 2. Safety inadequacy in each field as per Questionary Survey.

Construction Field	Question no.	Question.	RRI
Jobsite general	JG9	Is complaint register managed and checked daily?	0.7778
Electrical	EL8	Do you check that electrical cables and leads are protected from damage?	0.7156
Excavation/Earthwork	Ex/Ea9	Do you ensure that barriers or other protection is there to stop people/vehicles falling in excavated pit?	0.7244
Safety netting	SN4	Is there a rescue/recovery plan in place?	0.7244
Fire risk & management	FR2	Do you ensure that all staff knows what these procedures are?	0.6889
Fire prevention	FP7	Is fire engine access road provided in case if fire is caught?	0.6800
Flammable liquid/ materials	FL1	Empty containers removed?	0.7378
First Aid	FA3	Provision for Signs directing persons to the site of the first aid kit?	0.7600
Housekeeping & sanitation	HK2	Do you ensure debris is not allowed to be collected around the site?	0.7333
Lifting operations	LO2	Do you check that the rated capacity/safe working load is clearly marked on the lift?	0.8889
Hoists, Crane & Derricks	HC7	Alarms working and audible?	0.7333
Personal protective equipment	PPE6	Do you ensure the footwear is manufactured to CE standards?	0.7200
Road works	RW1	Are all road signs properly secured ?	0.6756
Construction areas- secured access/after hours	CA5	Are id badges given to workers and visitors to access construction sites?	0.6622
Scaffoldings/Working platform	SW7	Do you ensure that there are sufficient warnings notices in place to stop any person using an incomplete scaffold?	0.7111
Site set-up	SS7	No smoking signs?	0.6889
Traffic management	TM1	Has one way or turning points been identified?	0.7556
Working at heights	TM8	Do you ensure people are kept away from areas below the roof work?	0.6978
Ladders	L6	Ladders are inspected and in good condition?	0.6933
Welding & cuttings	WC5	Gas lines and power cables protected and in good condition?	0.7200
Concrete construction	CC3	Caps on rebar	0.6444

Improved construction safety and planning lets a developer to effortlessly run the construction projects and prevent any accidents in the course of construction. Consequently, the member of staff, who hires the workers and is responsible for the safety of these employees, can get peace of mind. Moreover, it is not only the employees who are at stake, due to risks such as collapse of a heavy equipment, there can be numerous falling effects on the work sequence. Due to an machinery failure, the task for which this machine is used, gets postponed, as the outcome of which the entire procedure gets postponed and the project schedules will be changed. Equipment failures which can undergo midway, during the construction phase, there are risks that raw material of that specific batch is squandered. The bottom line gets affected by all these factors eventually.

Autodesk Revit is building information modelling - BIM software for designers, contractors, MEP engineers, structural engineers and architects developed by Autodesk. It permits operators to design a building and design the building and all its elements in 3D, use 2D drafting elements to interpret the 3D model, and utilize the database of the building model to access thru building information data. Revit architecture is 4D capable BIM tool to develop and track different stages in the life-cycle of the structure, from conception to construction stage and later demolition and/or maintenance. Revit software was envisioned to allow architects, engineers and other building professionals to plan and document a building by generating a parametric three-dimensional model that comprises both the non-geometric and geometry design and construction data, which is also called as BIM or Building Information Modelling. Revit is utilized a very powerful collaboration tool between different disciplines.

With utilization of three-dimensional, on-going, dynamic building modelling software like Revit architecture, the engineer or architect is offered the chance to plan a model and its elements in 3D, with 2D drafting elements interpret the model, and building model data-archive helps access the building information. Revit's 4D abilities allow for tracking and planning stages of a building's lifecycle and material quantity calculations part. While the designer uses Revit architecture as an early design tool, the 3-D massing of a building can be swiftly understood and developed.

One of the important duties in site management is the process of site layout planning. Massive cost overruns and time waste could cause in huge projects, if effective site planning is not implemented. A thorough planning of the site location and site layout of interim facilities can bring about development by increasing worker morale and lessening travel time, waiting time all by showing improved and safer work environment. One of the most overlooked facets in the construction industry is site planning. It is crucial to recognize that the site policies put in place will become rules and terms that site employees will reside along with for the entire duration of the construction phase. Therefore, it is imperative for very careful pre-planning. Substantial saving can be ensued if the labour force moved quickly and liberally within the site.

A Real-time project of the construction site including the building plan in constructing the virtual building model using Revit architecture with the jobsite work plan layout having stated its advantages in a better safety management system, this study is aimed at proposing the solutions using BIM visualization for the crucial factors. In high-rise building the 3D jobsite workplan layout model of under-construction projects was planned using Revit Architecture. The entire construction site layout was split into different parts like electrical maintenance area, heavy equipment parking area, batching plant, material storage area, excavation area, bar bending area, segregation area, welding & cutting area and PPE training area are shown in the model. Along with this site office, assembly point and first aid room are shown in the model. And based on different areas, general guidelines are given to improve safety at construction site.

4. Conclusions

The following are the conclusions drawn from RRI analysis and BIM Visualization adoption for a better construction safety management system:

- It was thus derived from the responses and through analysis that the most neglected safety precaution taking place in construction industry is open ended rebars i.e., rebar caps are not being utilized on sites.
- The second most significant factor is ID badges not being issued foe labourers and visitors to site. This is an significant disregard to safety because if an pedestrian unknowingly enters and construction site and is not aware of the site safety guidelines may come into danger. Also not recoding and movement of people entering and exiting site may lead to cases of theft and any accidents or unfortunate incidents taking place may fall under the responsibility of Contractor or Client.
- Improper or no CCTV monitoring on construction site is again another major reason which is contributing to inadequate safety monitoring on construction site. It is very important to have CCTV monitoring on every construction site regardless of the scale of project.
- Although safety signs and road signs are placed in the construction site, in most cases they are not secured properly. This is especially important for road signs as they help avoiding any traffic related incidents.
- Segregation of waste is another important part of construction safety as well as for environment. Through the survey its understood that combustible and non-combustible waste are being disposed into same bin.
- All construction site needs to have a fire engine access road, although there are roadways which allow access for heavy vehicles these road-way do not access all parts of an construction site. Proper precaution is needs to make sure a fire access road is made available as freely as possible.
- It was found that although safety signs and road signs are put up, no smoking signs are insufficient in numbers. This is very important as many flammable liquids and materials are utilized throughout the construction industry.
- In many construction sites it was discovered through the survey that although hazardous waste is being generated no proper policy or procedure is put in place for handling of such waste.
- Although most construction sites have fire risk assessment and have procedure for evacuation in case of fire break out, from the survey its been noted that not all staff and labourers are trained for it.
- Within construction site ladders are used for multiple jobs within the construction activities. Although ladders are in use they are not properly inspected, this is a major problem for construction safety as falling from heights is one of the most common accidents taking place in construction site.

BIM Visualization is going to play a very important role in the future of both the construction industry as well as construction safety. It as has showed to be very effective in planning for construction safety from the initiation stages of the project itself. Various aspects of construction safety can be defined in 3D and planned out to make minimise risks and hazards as effectively as possible. Also, it allows a designer to plan as per the safety by-laws so it can be applied affectively.

References

[1] Sawacha, E., Naoum, S. and Fong, D., 1999. Factors affecting safety performance on construction sites. *International journal of project management*, *17*(5), pp.309-315.

[2] Mohamed, S., 1999. Empirical investigation of construction safety management activities and performance in Australia. *Safety Science*, *33*(3), pp.129-142.

[3] Gibb, A., Haslam, R., Gyi, D., Hide, S. and Duff, R., 2006, November. What causes accidents?. In *Proceedings of The Institution of Civil Engineers-Civil Engineering* (Vol. 159, No. 6, pp. 46-50). Thomas Telford Ltd.

[4] Guha, H. and Biswas, P.P., 2013. Measuring construction site safety in Kolkata, India. *International Journal of Scientific and Engineering Research*, *4*(5), pp.2138-2143.

[5] Singh, K., 2014. Safety in Indian construction. *International Journal of Engineering Research & Technology*, pp.1564-1566.

[6] Sijie Zhang, Kristiina Sulankivi, "BIM-based fall hazard identification and prevention in construction safety planning" Elsevier Safety Science 72 (2015) 31–45, February- 2015.

[7] Construction, M.H., 2013. Safety management in the construction industry: Identifying risks and reducing accidents to improve site productivity and project ROI. *Bedford, MA: Author.*

[8] Marefat, A., Toosi, H. and Hasankhanlo, R.M., 2018. A BIM approach for construction safety: applications, barriers and solutions. *Engineering, Construction and Architectural Management.*

[9] Hossain, M.A., Abbott, E.L., Chua, D.K., Nguyen, T.Q. and Goh, Y.M., 2018. Design-for-safety knowledge library for BIM-integrated safety risk reviews. *Automation in Construction*, *94*, pp.290-302.

[10] Akram, R., Thaheem, M.J., Nasir, A.R., Ali, T.H. and Khan, S., 2019. Exploring the role of building information modeling in construction safety through science mapping. *Safety Science*, *120*, pp.456-470.

[11] Liu, D., Jin, Z. and Gambatese, J., 2020. Scenarios for integrating IPS–IMU system with BIM technology in construction safety control. *Practice Periodical on Structural Design and Construction*, *25*(1), p.05019007.

[12] Lu, Y., Gong, P., Tang, Y., Sun, S. and Li, Q., 2021. BIM-integrated construction safety risk assessment at the design stage of building projects. *Automation in Construction*, *124*, p.103553.

[13] Akinlolu, M., Haupt, T.C., Edwards, D.J. and Simpeh, F., 2020. A bibliometric review of the status and emerging research trends in construction safety management technologies. *International Journal of Construction Management*, pp.1-13.

[14] Mayer, P., Funtík, T., Gašparík, J. and Makýš, P., 2021. Analysis of the Current State of Automation of Hazard Detection Processes in BIM in Slovakia. *Applied Sciences*, *11*(17), p.8130.

[15] Luo, F., Li, R.Y.M., Crabbe, M.J.C. and Pu, R., 2022. Economic development and construction safety research: A bibliometrics approach. *Safety science*, *145*, p.105519.

Numerical Study on V Shaped Natural Draft Cooling Tower Supporting System

Harish Sagar M[1], Mehar Babu[2], Umesh S S[3], D.S. Ananda[4]

[1,2] Assistant Professor, Department of Civil Engineering, REVA University

[3] Professor, Department of Civil Engineering, Sayadri College of Engineering

[4] Chief Engineer at KPCL

(harishsagar.m@reva.edu.in meharbabu.ravula@reva.edu.in)

ABSTRACT. Cooling towers generally used to extract waste heat to the atmosphere from cooling water stream to a lower temperature and hence it rejects heat from water. The size of the cooling tower is large, with thin wall and complex geometry. Wind load plays a dominant role in the design of cooling tower. These cooling towers are subjected to self-weight and dynamic loads such as wind and earthquake. This paper presents the study of cooling tower of height 200m above the ground level. The Analysis of a cooling tower is done for a wind load using a finite element analysis considering V frame shell base supporting system. The analysis is performed in Commercial software STAAD PRO and the results obtained are presented in this paper.

Keywords: Cooling Tower, dynamic, Earthquake, V Frame, FEA

1. Introduction

Hyperbolic cooling tower consists of a thin and large shell reinforced cement concrete structure and which protects the environment. The natural draft cooling tower in a thermal power plant they play a very important rule and also a essential component, they also used in nuclear power plants. In thermal power plants the water which is converted in to a steam drives the turbines from which it develops an electric energy. Then after the steam which is drived the turbine is made condensed and again passed to the boilers were the hot water again converted to a steam jet. The warm air which is present in the cooling tower and the cooled air which is present outside the cooling tower develops a difference in density and creates the natural draft in the interior, this causes flow of warm air upward which forms a continuous steam of fresh air from the air inlets to the tower, the cooling tower shell which is supported on truss or column frame works which has space in between a pair of column causes the air inlet to the tower.

There are also dry cooling towers it evaporates as a convective cooling. In this case of dry cooling tower water distribution, water basin and fill are replaced with a piping system which is closed around the air inlet. Approximate of 30% thermal efficiency for dry cooling towers then compare to wet cooling tower. If in case the flue gas is cleaned with a washing technology then it will be discharged to the atmosphere by a cooling tower with a upward flow this causes a save of reheated cleaned flue gas and construction of smock stack.

It indicates that initially the cooled water is taken in to the water basin, the cooled water it is passed to a condenser so that water get cooled, if in case of hot water. Then it is passed to steam circuit in this chamber the water is converted in to a steam jet, the steam water in the form of a jet is passed in turbo generator in which it drives the turbine and creates a electric energy, which is the form of thermal energy to mechanical energy and then mechanical energy to electrical energy, this steam jet is again passed in to a condenser in which steam is converted in to hot water and then this hot water is distributed evenly from the channels and pipes above a fill in the cooling tower, as water drops down comes in contact with the rising cooler air and finally the cooled water drops in to a water basin. The process which takes place in boiler to cool water is as follows, the coal is ignited in boiler with high temperature and then the waste gas which comes in contact with air is passed in to chimneys so that it should not harm to the environment and cause the environment pollution.

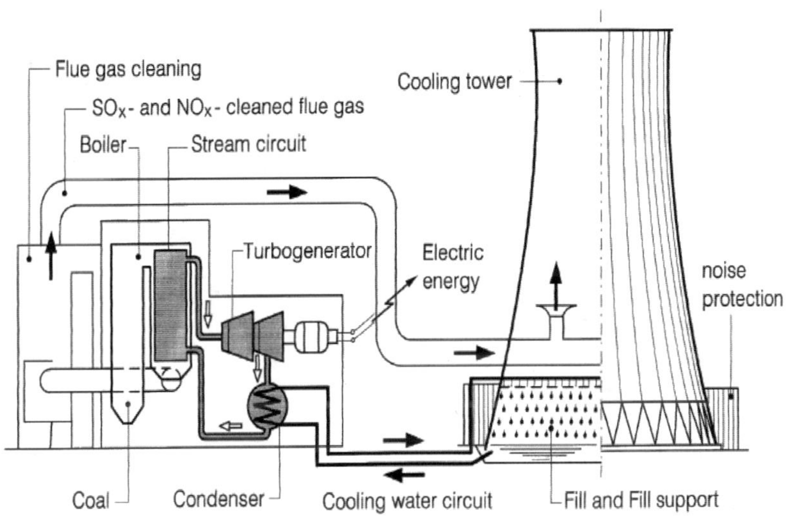

Fig. 1. Mechanism of cooling tower

2. Literature Review

Aswath M.U, Veena N, Sayeed Sulaiman [1] This paper presented the numerical analysis of natural draft cooling tower of a column support under a wind load and seismic load i.e dynamic loading. In this case of numerical analyses, two different types of the supporting column frame systems are adopted and the dynamic response of NDCT cooling tower is examined. In case of a static load (for self-weight), maximum deflection and principle stress is greater for A-frame column support then H-frame column support. In case of wind load, the principal stress and maximum deflection is greater for H-frame column support then A-frame column support. In case of seismic load, maximum principal stress is greater for A-frame column support and the maximum deflection is greater for H-frame column support.

Kopenetz Ludovic Gheorghe, Catarig Alexandru [2] In this case the paper presents that the numerical examples and the theoretical background and the beam models used are as presented follows. The modelling of cooling tower in this case is done by using as a bar type finite element analysis which is having three or four noded plate shells which offers a good vibration for analysis and gives a accuracy for the engineering needs in practical. This present free vibration analysis used in seismic analysis with the enforced seismic design needs.

V.S.N Raju [3] In this paper he investigated that MSC/NASTRAN software's which are very convenient for the analysis of cooling towers because it gives the advantage to account for the curvature of surfaces by using quad8 elements. And this same can't be done in case of sap IV.In this sap IVonly plane surfaces are considered. And the result which is obtained through MSC/NASTRAN with quad8 elements are quite reliable for cooling towers as compared to the software package sap IVparticularly on the bases of the rackers columns are joining, as the quad8 elements will give the forces at four corners and also at the c.g of the elements. At the base of the cooling towers it will going to cover the possible stress concentration.

Shailesh s.Angalekar[4] In this paper analysis of natural draft cooling tower is done by using FEMA Using equivalent plate method concept, A cooling tower of single case is taken for both I and V Supporting systems. It is demonstrated on the basis of behaviour of equivalent plates are identical to the behaviour where the columns with actual supports are considered. For this the wind loads over the structure.

3. Finite Element Analysis

In the thermal power plants the cooling tower which are present it consists of three or four node plates of shells, and the shape of the cooling tower is hyperbolic and at the bottom portion of the cooling tower which is conical.

The RCC shells of cooling tower are supported by 88 pairs of diagonal RCC raker columns the shells and the raker columns are joined with the help of ring beams. At the place where columns are provided the open system will be created in between of the paired raker columns and this is called as air inlet opening. The hot water which is cooled in a natural draft cooling tower are retained in the pond wall. When the foundation are provide at the bottom of the cooling tower in that case a ring shaped RCC horizontal ring foundation is to be provided and this oil bearing capacity in case of ring foundation is taken to be 50 t/m2 and for the depth of 5m from the ground level.

3.1 Modelling Details

- Top dia. of cooling tower = 93 m
- Bottom dia. of cooling tower = 131.15 m
- Average dia. of cooling tower = 112.075 m
- Circumference of a cooling tower = 352.094 m
- No of divisions along the circumference of a circle = 82
- Height of a cooling tower = 188 m
- No of divisions along the height of a circle = 60
- Average Width of the shell = 4.29 m.
- Average height of the shell = 3.13 m.
- Aspect ratio mainly depends upon the average shell width and average shell height of the cooling tower. In order to get accurate analysis, result the Aspect ratio should be with in 1.5. Aspect ratio = 1.37

3.2 Wind Loads as per IS875 Part-3

In this case the wind load is calculated as per external pressure distributions in IS: 875 (part3) – 1987, Table No. 22 [5]. Distribution of wind pressure on the circumference of the cooling tower from the outside of the shell is unsymmetrical along the circumference and also along the meridian height along the direction of the wind. The distribution of wind pressure on the circumference of cooling tower shell is defined as in Fig. 2 and the calculation of the wind pressure along the height is in Table 1.

IOP Conf. Series: Materials Science and Engineering 1255 (2022) 012007 doi:10.1088/1757-899X/1255/1/012007

IS : 875 (Part 3) - 1987

TABLE 22 EXTERNAL PRESSURE DISTRIBUTION COEFFICIENTS AROUND SPHERICAL STRUCTURES
(Clause 6.2.2.13)

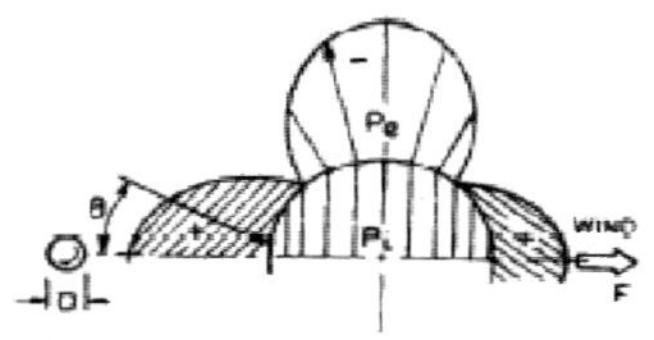

Position of Periphery, θ in Degrees	C_{pe}	Remarks
0	+1·0	C_f = 0·5 for DV_d < 7
15	+0·9	= 0·2 for DV_d ≥ 7
30	+0·5	
45	−0·1	
60	−0·7	
75	−1·1	
90	−1·2	
105	−1·0	
120	−0·6	
135	−0·2	
150	+0·1	
165	+0·3	
180	+0·4	

Fig. 2. Circumferential net wind load pressure distribution as per IS 875 [5]

Table 1: Wind speed variation with height in different general buildings

COOLING TOWER FOR A HEIGHT OF 188 METER						
Pressure for general buildings						
$V_z=V_b*K_1*K_2*K_3$ (m/s)						
Height	V_b (m/s)	K_1	K_2	K_3	V_z(m/s)	P_z(kN/m^2)
10	39	1.05	0.91	1	37.2645	0.83318578
30	39	1.05	0.99	1	39.3232	0.9875212
20	39	1.05	1.01	1	41.3595	1.02636494
40	39	1.05	1.09	1	44.6355	1.19539672
50	39	1.05	1.12	1	45.864	1.2621039
60	39	1.05	1.136	1	46.5192	1.29842158
70	39	1.05	1.152	1	47.1744	1.33525441
80	39	1.05	1.168	1	47.8296	1.37260238
90	39	1.05	1.184	1	48.4848	1.4104655
100	39	1.05	1.2	1	49.14	1.44884376
110	39	1.05	1.208	1	49.4676	1.46822607
120	39	1.05	1.216	1	49.7952	1.48773717
130	39	1.05	1.224	1	50.1228	1.50737705

140	39	1.05	1.232	1	50.4504	1.52714572
150	39	1.05	1.24	1	50.778	1.54704317
160	39	1.05	1.246	1	51.0237	1.56205078
170	39	1.05	1.252	1	51.2694	1.57713083
180	39	1.05	1.258	1	51.5151	1.59228332
190	39	1.05	1.264	1	51.7608	1.60750825
200	39	1.05	1.27	1	52.0065	1.62280563

3.3 Geometry of Natural Draft Cooling Tower (NDCT) steps followed in STAAD PRO

Step 1. New project > space > open structural wizard > surface/plate models Fig. 3.

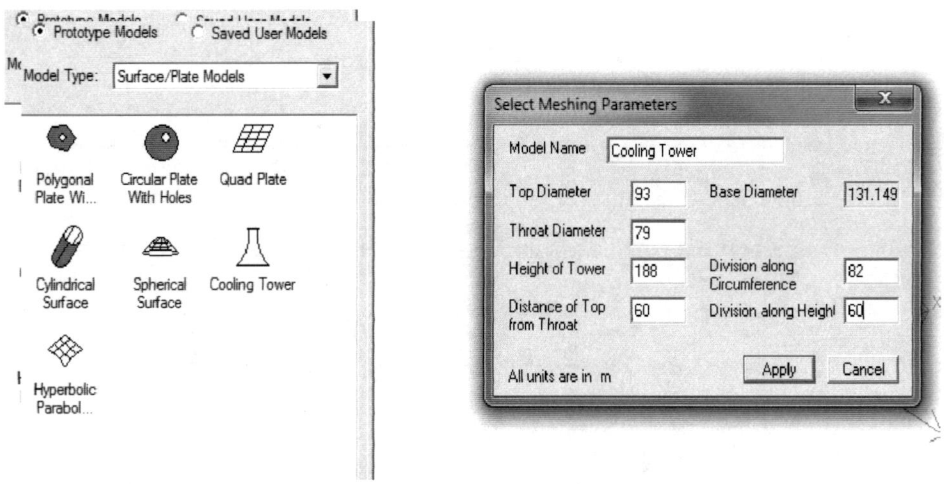

Fig. 3: Geometry of NDCT

After providing the data as in Fig.3 the cooling tower of height 188 m model shape is generated in software STAAD-Pro as shown in Fig. 4.

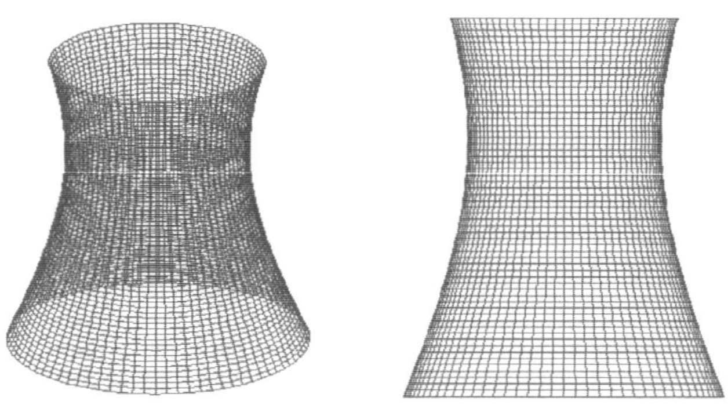

Fig. 4. NDCT modelled in STAAD Pro

Step2. This model does not contain ring beams and raker columns and hence in order to provide ring beams and raker columns, the supporting shell plates are to be provided these shell plates are of triangular and quadrilateral and this supporting shell plate nodes are created with the help of interpolation with same slopes as geometry of cooling tower by proper calculation. The variation of slope in between supporting shell plates and geometry of cooling towers should not be occurred. The orientation of ring column with raker columns is kept perfect and the load transfer followed proper path through the nodes using the interpolation of nodes that is shown in Fig. 5. Every single node of ring beam is kept coincided with the raker columns so that load transfer is perfectly through node to node.

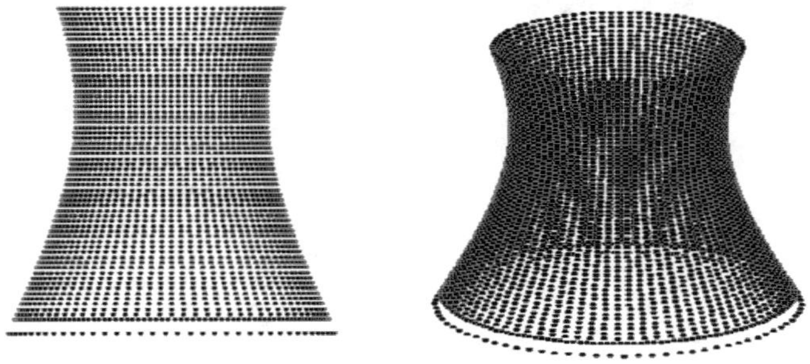

Fig. 5. Interpolation of nodes

Ring beams and raker columns are provided to supporting shell plates. The length, height and diameter of ring beam and raker column is kept same. The Ring beams and raker columns of the cooling tower base is shown in Fig. 6 from the STAAD Pro modelling and Final model of the cooling tower showing ring beam and raker column is shown in Fig. 7.

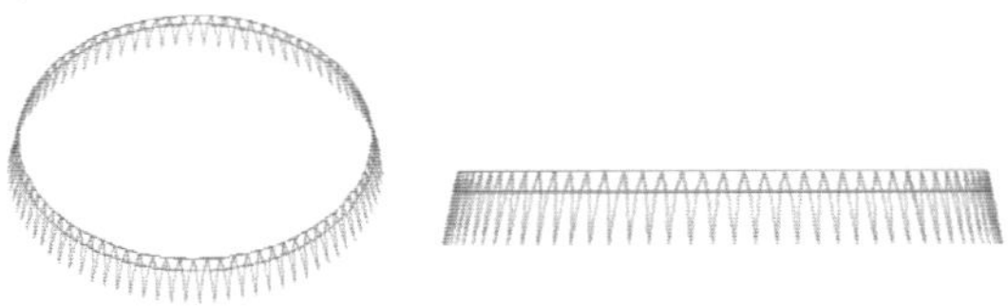

Fig. 6. Ring beams and raker columns

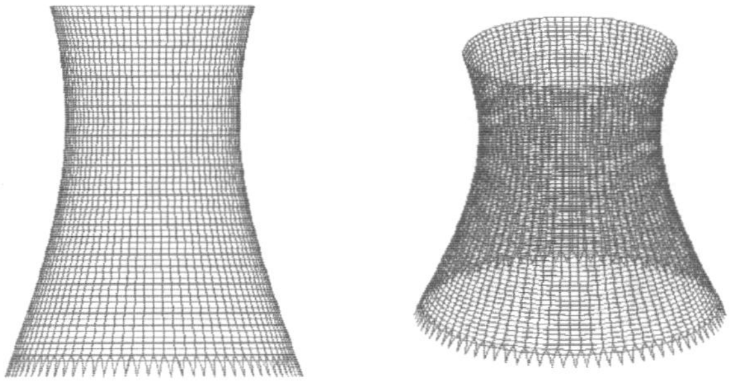

Fig. 7. Final model after providing ring beam and raker column

3.4 Property of the Cooling Tower

The plate thickness is given to the shell elements of the cooling tower model and the physical property of the whole structure with varying shell thickness is defined in properties tab of STAAD Pro as shown in Fig. 8. The material assigned is Concrete of grade M35.

IOP Conf. Series: Materials Science and Engineering 1255 (2022) 012007 doi:10.1088/1757-899X/1255/1/012007

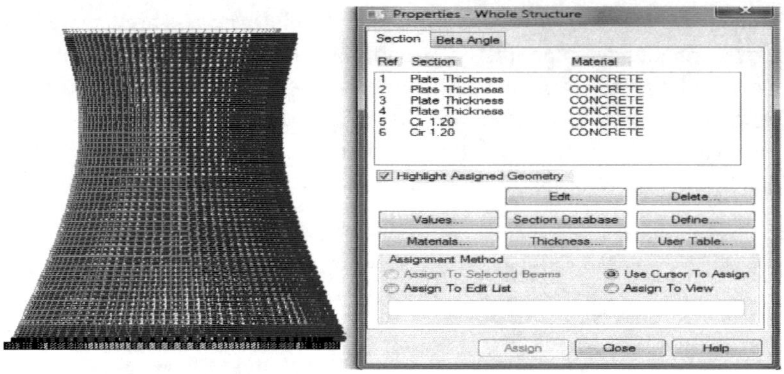

Fig. 8. Properties of cooling tower

The thickness of the cooling tower shells varies from bottom to top. The cooling tower shells are divided in to four strips as follows.

1. Up to height of 0 to 20m from base, the thickness of shell is 1m and the diameter of cooling tower varies from 131.08 to 117.33m.
2. From 20m to 40m from base, the thickness of shell is 0.8m and the diameter of cooling tower varies from 117.33 to 106.47m.
3. From 40m to 100m from base, the thickness of shell is 0.6m and the diameter of cooling tower varies from 106.47 to 82.19m.
4. From 100m to 188m from base, the thickness of shell is 0.4m and the diameter of cooling tower varies from 82.19 to 93m.

The ring beam provided to the cooling tower shells has diameter 1.2m and its length is 2.5m. The raker columns provided to the cooling tower has diameter 1.2m and its height 9m.

3.5 Load Combinations considered for Analysis

All the components of cooling tower structure are designed for both working stress and limit state method, excluding the tower elements like fill supporting for a structure, staircase that are preferred and a platform. The Load Combinations considered for the analysis are as follows

a. Design of cooling tower for Working stress method
   i) DL + WL
   ii) DL + 1/3(WL) + EQ
b. Design of cooling tower for limit state method
   i) 1.5 (DL) + 1.5 (WL)
   ii) 0.9 (DL) +1.5 (WL)
   iii) 1.5 (DL) + 1.5 (SLX) + 0.45 (SLZ)
   iv) 0.9 (DL) + 1.5 (SLX) + 0.15 (SLZ)

Where DL = Dead Load, WL = Wind Load, EQ = Earthquake Load

,

Fig. 9. V-Column

3.6 MESHING

The cooling tower structure is modelled with the help of beam and plate elements in the commercial software STAAD-Pro v8i. The shells which are used in cooling tower for meshing is quadratic 4-noded plate element, 3D beam is done for modelling of raker columns. At the base of the ring beam 3-noded triangular plate elements are used. The total number of nodes used is 5166, the total number of beams used is 328 and total number of plates 5002 in the entire model.

Fig. 10. V type Column supports for a cooling tower with Shell and mesh

4. Results and Discussion

V-Column supporting system of cooling tower is analyzed by a finite element analysis. Wind load is calculated as per IS: 875 part-3 code [5] for open terrain category. The results for circumferential bending moment and forces are taken from the Wind analysis of the V-Column supported cooling tower. A typical view of the deformed shape of a cooling tower is shown in Fig. 11 and the deformed view at the throat is shown in Fig. 12. The bottom of V-Column is considered as fixed and wind is applied on the whole structure.

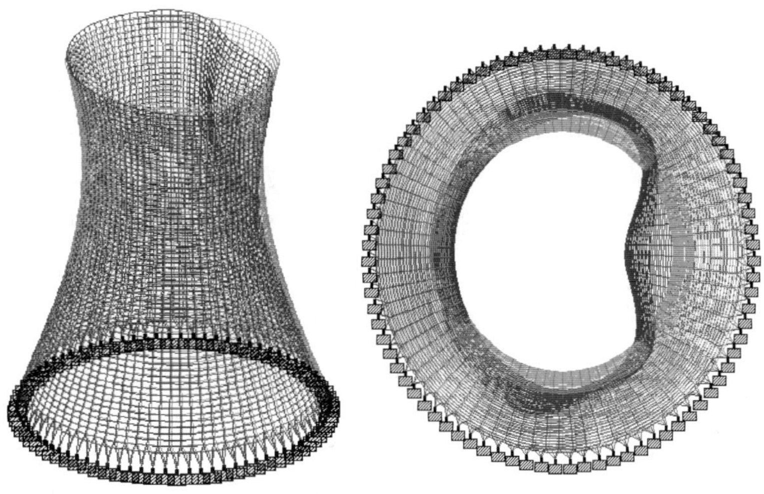

Fig. 11. Deformed shape of cooling tower

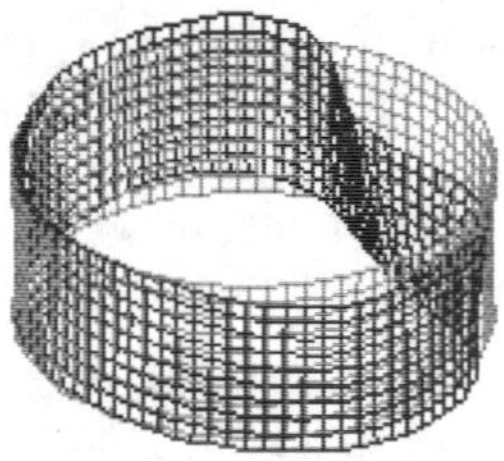

Fig. 12. V-col at throat

From the analysis of V-frame supporting system Cooling Tower, the Circumferential Moment is extracted at bottom level of a cooling tower for windward direction and Leeward direction is plotted along the circumferential angle from 0-360° shown in Fig. 13. At windward direction, the Circumferential Moment increases gradually from 0-90° and decreases gradually from 90-180°. On the windward direction, Maximum Circumferential Moment is observed at 90° circumferential angle as surface is perpendicular to the wind load. Drastic variation in Circumferential Moment is observed on windward direction whereas uniform redistribution of Circumferential Moment is observed at leeward direction. Hence Uniform redistribution of circumferential stresses is observed at leeward side. At a circumferential angle of 90°, the circumferential tensile stress is lowest compared to all other circumferential angles, due to maximum circumferential Moment.

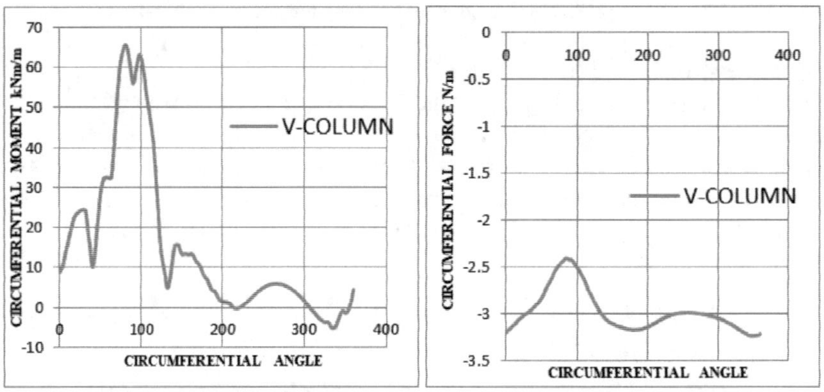

Fig. 13. (a) Plot of circumferential moment, (b) Plot of circumferential force at bottom (ring beam)

The Circumferential Moment at throat and top is observed to follow same pattern with maximum at 90° and uniform redistribution at circumferential angle beyond 180° i.e the leeward direction. The magnitude of Circumferential Moment and Circumferential force is very much negligible compared with bottom ring beam as shown in Fig. 14, Fig. 15.

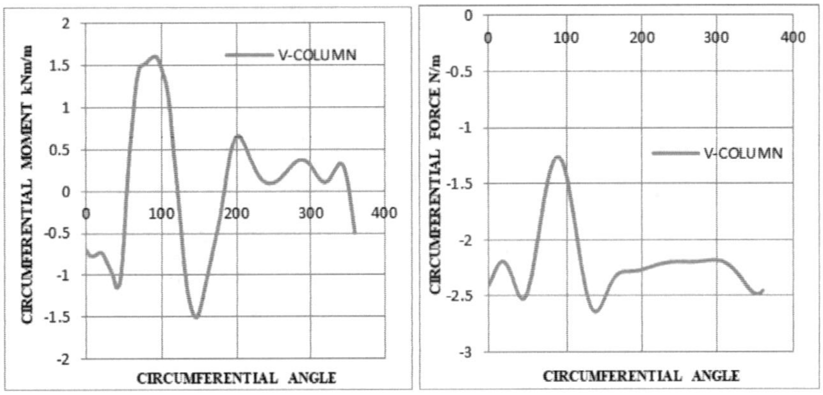

Fig. 14. (a) Plot of circumferential Moment, (b) Plot of circumferential force at throat

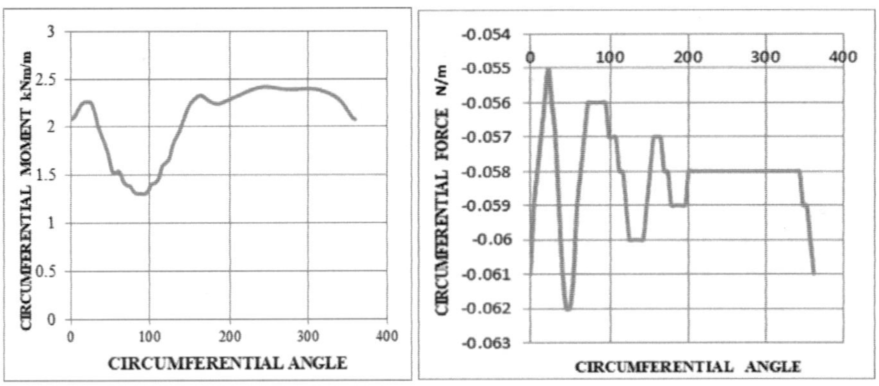

Fig. 15. (a) Plot of circumferential Moment, (b) Plot of circumferential force at top

5. Conclusions

The circumferential moment showed increasing pattern. At windward direction, the Circumferential Moment at bottom ring beam increased gradually from 0-90° and decreased gradually from 90-180°. On the windward direction, Circumferential Moment at bottom ring beam is observed maximum at 90° circumferential angle as surface is perpendicular to the wind load. At bottom ring beam, drastic variation in Circumferential Moment is observed on windward direction whereas uniform redistribution of Circumferential Moment is observed at leeward direction. Hence Uniform redistribution of circumferential forces is obtained at leeward side. At throat level and at top level, the Circumferential Moment is observed to follow same pattern with maximum at 90° and uniform redistribution at circumferential angle beyond 180° i.e the leeward direction. The magnitude of Circumferential Moment and Circumferential forces are negligible compared with bottom ring beam.

1. At windward direction, the Circumferential Moment initially decreases and then increases gradually from 30-90° and again decreases from 90 to 135°.
2. On the windward direction, Circumferential Moment and force at bottom ring beam is observed maximum at 90° circumferential angle as surface is perpendicular to the wind load.
3. At bottom ring beam, gradual variation in Circumferential Moment is observed on windward direction whereas uniform redistribution of Circumferential Moment is observed at leeward direction.
4. At throat level and at bottom level, the Circumferential Moment is observed to follow same pattern with maximum at 90° and uniform redistribution at circumferential angle beyond 180° i.e the leeward direction.

IOP Conf. Series: Materials Science and Engineering 1255 (2022) 012007 doi:10.1088/1757-899X/1255/1/012007

5. The magnitude of circumferential moment and circumferential force at throat are less compared with bottom ring beam.

6. REFERENCES

[1] Veena N, Sayeed Sulaiman, Ashwath MU, (2013) "Comparative study of the effect of seismic and wind load with A-frame and H-frame column support", International Journal of Science and Technology, issue 8, vol. 1, ISSN 2321-919X.

[2] Kopenetzludoricgheorghe, Catarig Alexander., (2011) "Practically structure analysis of large cooling tower", Journal of Applied Engineering Science, Vol. 1(14), Issue 4, ISSN 2247-3769.

[3] Raju VSN, (2007) "Analysis of hyperbolic shell natural draught cooling towers using MSC/NASTRAN".

[4] Shailesh S. Angalekar, Kulkarni AB, "Analysis of Natural Draught Hyperbolic Cooling Towers by Finite Element Method Using Equivalent Plate Concept", International Journal of Engineering Research and Applications (IJERA), ISSN: 2248-9622, Vol. 1, Issue 2, pp.144-148.

[5] IS: 875 (Part 3): 1987, "Code of practice for design loads (Other than earthquake) for buildings and structures", Bureau of Indian standards, New Delhi.

[6] IS 11504-1985 (Reaffirmed 1992), "Criteria for structural design of reinforced concrete natural draught cooling towers", Bureau of Indian Standards, New Delhi.

[7] IS 1893 (Part 4):2005, "Criteria for earthquake resistant design of structures", Bureau of Indian Standards, New Delhi.

IOP Conf. Series: Materials Science and Engineering 1255 (2022) 012008 doi:10.1088/1757-899X/1255/1/012008

Seismic Analysis of Diagrid Structure on Sloping Ground

Shylaja N[1], Ashwini R M[2] and Babu E R[3]

[1] Assisatnt Professor, School of Civil Engineering, REVA University, Bengaluru-560064
[2] Assistant professor, GITAM School of Technology , GITAM University,Bengaluru-561203
[3] Associate Professor, Bangalore Institute of Technology, Bengaluru-560004
Corresponding Author: nshylak@gmail.com

Abstract: Lateral efficiency of a building is very important in high-rise buildings, when it comes tobuilding them on sloping ground it is highly required to design the structure for wind load and earthquake loads. The diagrid structure thus provides very high lateral stability and reduces the lateral displacement of the building. In recent years the diagrid structure has attracted many researchers and scholars due to this greater structural efficiency. Diagrid structural system has evolved to be structurally the most efficient, aesthetically pleasant and highly economical. The present paper involves thestudy of diagrid structure on sloping ground and comparing it with the conventional system and finding the optimum angle of diagrid on different slopes of ground. A 12- storey building is used for comparing diagrid structure and conventional structure on sloping ground, a 48 storey building is used to find out the optimum angle of diagrid on sloping ground, ETABS software Is used for modelling and analysis, response spectrum method is used for analysis, the results show that diagrid structure is more efficient than conventional structure on sloping ground, and the optimum angle of diagrid lies between 60-75° on sloping ground.

Keywords: Diagrid structures, Sloping ground, Response spectrum method

1. Introduction

The scarcity of land on flat ground has led the people to construct buildings on hilly regions, in fact building on hilly areas provides additional advantages such as it provides excellent views, natural ventilation, natural drainage and additional space from lower floor levels, as a result the population in hilly regions is also increasing day by day, this rapid urbanization has led to the need for high-rise buildings. Designing a structure on sloping ground and flat ground are completely different as the mass and stiffness vary along the height and width of the building. Therefore, any high-rise structure on a sloping ground or even on a flat ground needs a special structural system [1]. As a structure's height rises, the system that resists lateral loads becomes more important than the structural system that resists gravity stresses. Shear wall, rigid-frame, wall-frame, outrigger system, braced tube system, and tubular system are the commonly utilized sideways load resisting systems.
The diagrid structural system has recently gained popularity. The term "diagrid" is formed by combining the words diagonal and grid refers to a single-thickness structural system that achieves structural coherence through triangulation. The triangulated member evolution is attained when the floor edge beams are connected to the grid, which accepts and deliver loads such as weight density and sideways force from several members and directions all over its configuration in an extremely dispersed manner. The diagrid system's structural strength and exquisite, unique appearance have helped to establish it as the most coveted tall building system of the new millennium. [2] illustrated a

paper which overview a stiffness based design approach for finding prefatory member sizes of steel diagrid structures in tall buildings. On the basis of several studies, it was suggested to incorporate a varied angle diagrid structure for a high rise building with an aspect ratio larger than 7 and a uniform angle diagrid for a tall building with its aspect ratio smaller than 7 as to save structural materials and, in turn, to design more defendable built environments. [3] Analysed model diagrid structures of 36-story. According to the outcome of analysis, the diagrid structures showed advanced strength with lower rigidity compared with the tubular structure. It was also noticed that as the slope of braces enhanced the shear lag effect enhanced and the sideway strength diminished. Both the strength and rigidity of diagrid structures enhanced significantly when the diagonal members were put back by buckling-restrained braces. The above-mentioned studies are carried on various aspects with respect to the architectural, aesthetical, angular aspects of the diagrid structure. They also involve the seismic effects that indulge in a diagrid structures behaviour [4-7]. The proper angle of inclination of the diagrid is studied and furthermore the complexity of the node construction has been emphasized [8]. Primarily, the various studies focus on strategies to enhance the performance of a building against lateral forces especially earthquake [9]. [10] studied the effect of mass irregularity at every storey one by one for 3 different zones (III, IV, V) and compared the Storey Drift and Storey displacement in different seismic zone. It is revealed from the literature that very less work has been done in the analysis of building on the sloping ground considering the combined effect of more than two irregularities. The diagrid system has attained favour amidst designers as a result of advancements in high-rise building construction, has emerged. Diagrid, a portmanteau of diagonal grid, is a framework of diagonally crossing components on the exterior of a structure that results in excellent structural performance in terms of resistant gravity and sideways loads while using reduced steel. The diagrid's fresh visual appeal has sparked renewed interest in architecture. Because the diagrid structure provides strong resistance to lateral stresses, it is an excellent choice for structures erected on sloping land, which are particularly vulnerable to earthquake and wind loads, and whose structures vary irregularly in horizontal and vertical planes.

2. Design and Analysis of 12 and 28 storey building.

2.1. Modelling of 12 Storey Diagrid Building with Conventional Building On Sloping Ground.

A Diagrid building on sloping ground and a conventional frame building on sloping ground is considered for analysis with 12 storey building with 24 m × 24 m plan dimension. The plan and elevation are shown in Fig 1.(a) to (d). with a storey height of the structure is 3.5m. The design dead and live loads are 3 kN/m^2 and 1.5 kN/m^2 on floor slab respectively. In diagrid structures, couple of braces is located on the verge of the building. The angle of bracing is selected by aspect ratio and the angle of inclination is made constant throughout the height. All the structural members are designed by using IS 800:2007. The Modelling, design and analysis of diagrid structure are executed using ETABS software. Seismic performance of the two models in terms of displacement of storey, drift, stiffness, Time period and total weight of the building is compared.

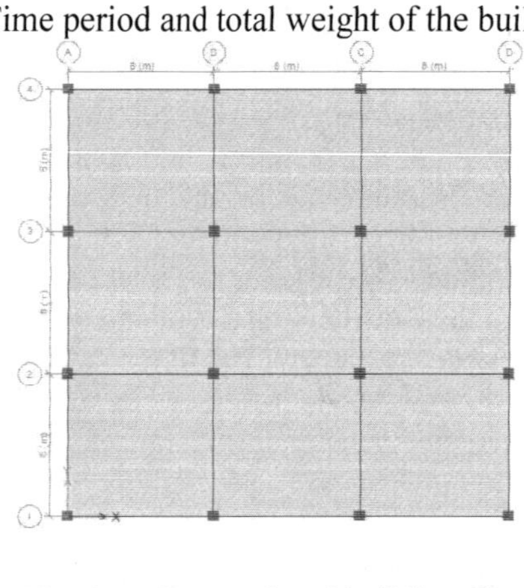

Fig. 1(a) Conventional building floor

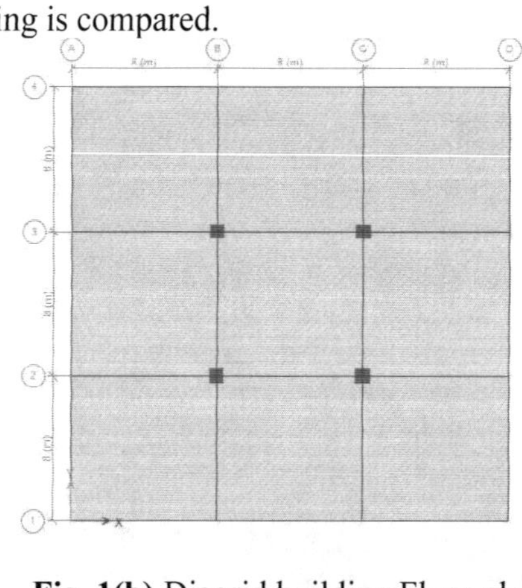

Fig. 1(b) Diagrid building Floor plan

IOP Conf. Series: Materials Science and Engineering 1255 (2022) 012008 doi:10.1088/1757-899X/1255/1/012008

plan

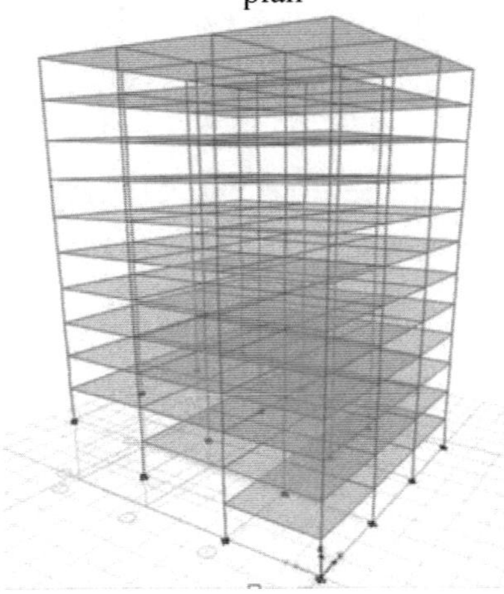

Fig. 1(c) 3D View of conventional building on slope 24°

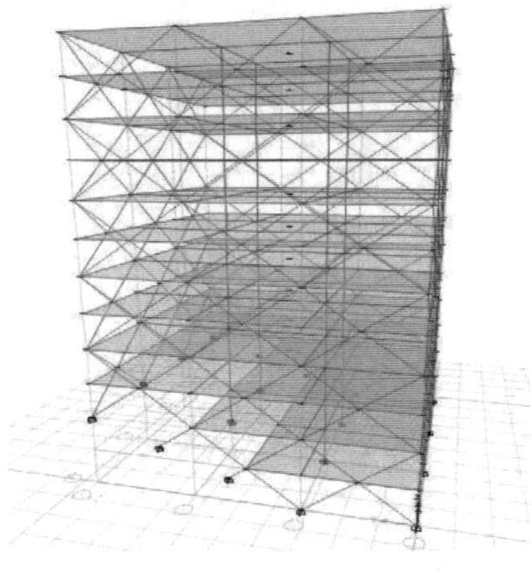

Fig. 1(d) 3DView of Diagrid Building on slope 24°

2.2. Modelling of 48 storey structures with different slopes of ground for different diagrid angles

A Diagrid building on sloping ground of 15° and 25° is considered for analysis with 48 storey building having 40 m × 40 m plan dimension. The typical plan is shown in Fig 2(a) and the elevation with different sloping ground and different diagrid angles is shown in Fig.2 (b) to (e) and Fig.3 (a) to (d) with a total height-172.8m. The design live load and dead loads on floor slab are 1.5 kN/m^2 and 4 kN/m^2 respectively.

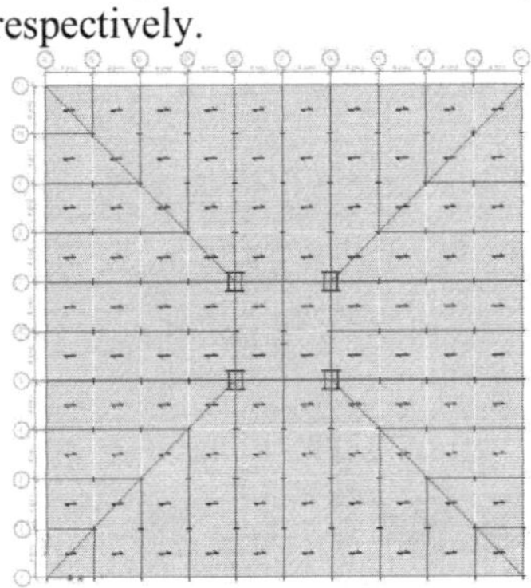

Fig.2(a) Floor Plan of all the 48 storey models

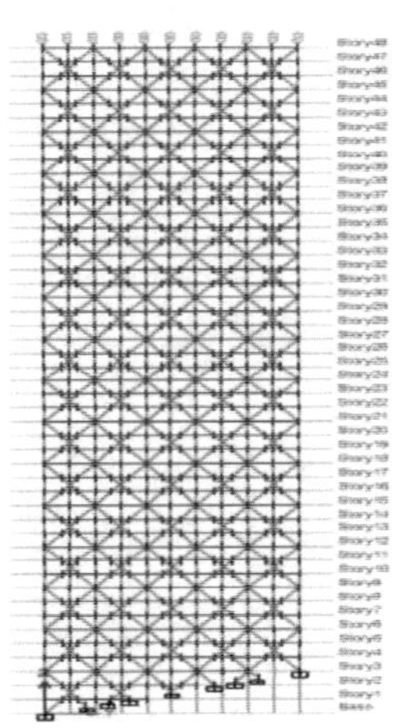

Fig. 2(b) Elevation of 48 storey structures for diagrid angles 53.47°

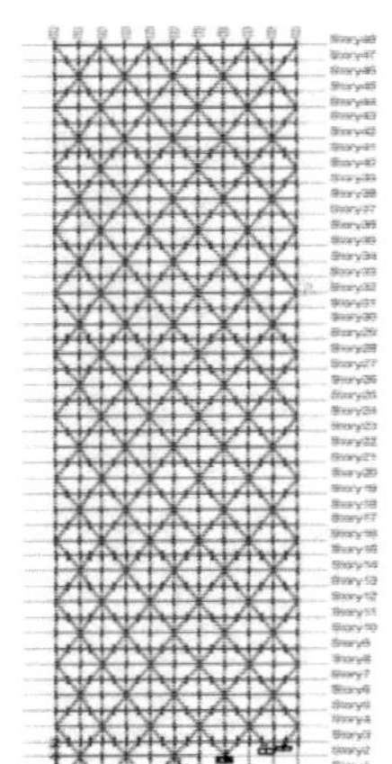

Fig. 2(c) Elevation of 48 storey structures for diagrid angles 60.94°

IOP Conf. Series: Materials Science and Engineering 1255 (2022) 012008 doi:10.1088/1757-899X/1255/1/012008

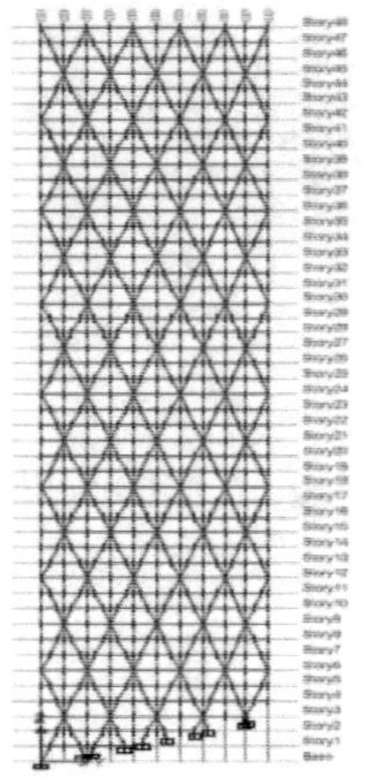

Fig 2(d) Elevation of 48 storey structures for diagrid angle 69.67°

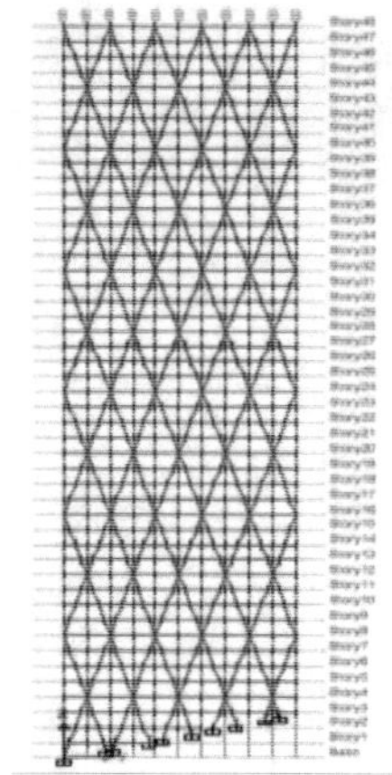

Fig. 2(e) Elevation of 48 storey structures for diagrid angle 74.47°

Fig. 2(f) Elevation of 48 storey structures for diagrid angle 79.5°

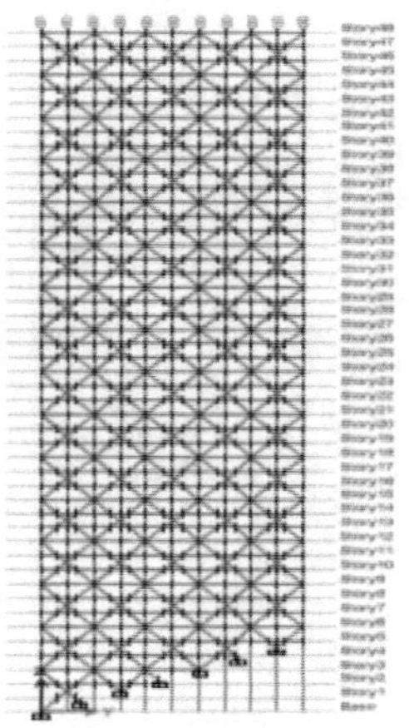

Fig. 3(a) Elevation of 48 storey structures for diagrid angle 53.47°

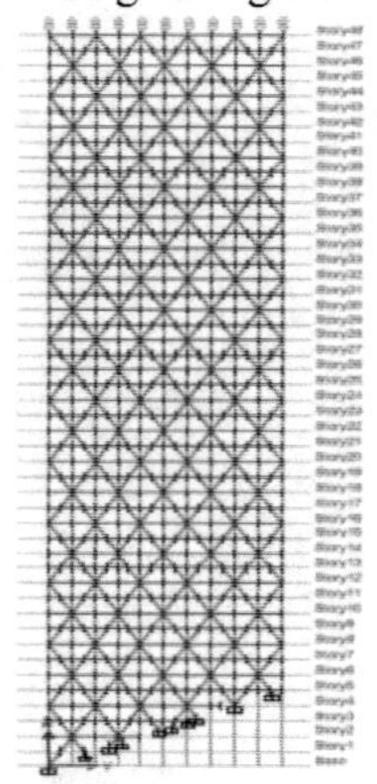

Fig. 3 (b) Elevation of 48 storey structures for diagrid angle 60.94°

Fig. 3 (c) Elevation of 48 storey structures for diagrid angle 69.67°

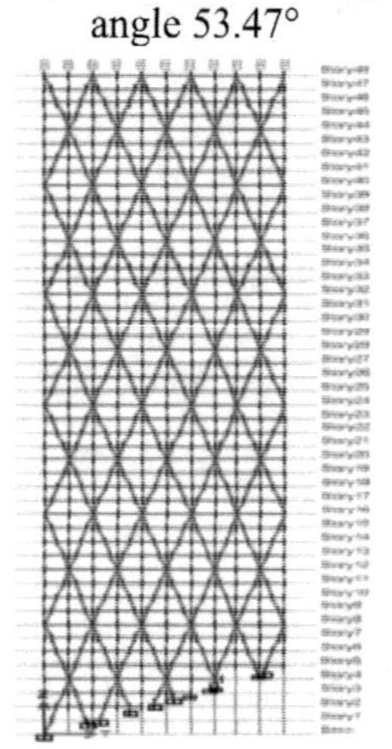

Fig. 3 (d) Elevation of 48 storey structures for diagrid angle 74.47°

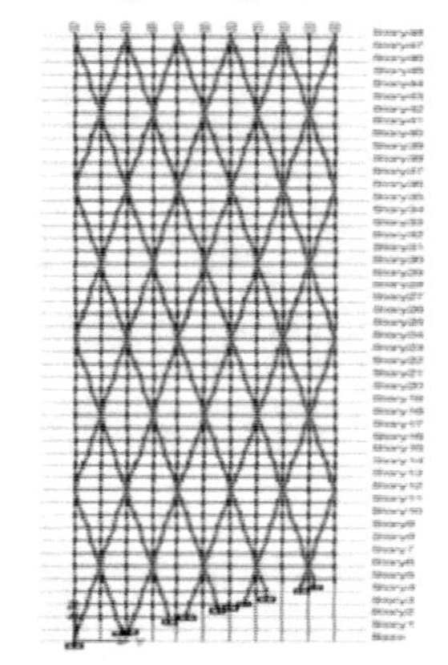

Fig. 3 (e) Elevation of 48 storey structures for diagrid angle 79.5°

2.3. Response spectral curve

The response spectrum approach of vibrational analysis has computational advantages for predicting deformations and internal forces in structural systems. Adopting smooth design spectra which were the average of numerous seismic motions, the method entails calculating just the greatest values of deformations and internal forces in each mode of vibration. The analysis of dynamic structure was done by two methods as Response Spectrum Analysis and Time History Analysis. The Response Spectrum Analysis method consists of evaluating the response in each mode of vibration and then superimposed the response in various modes to attain the total response. The vibrational analysis of several buildings is carried out by Response Spectrum Analysis as per IS 1893 (Part 1): 2016.

3. Results and Discussion

In all, 12 numbers of models were analysed by varying diagrid angles of 53.47°, 60.94°, 69.67°, 74.47° and 79.5° and response spectrum method was adopted for dynamic analysis, in seismic zones for type IV soil with the parameters of Base shear, Modal period, displacement, drift and Stiffness of the storey. The results were enumerated as follows.

3.1. Comparison of 12 storey conventional structure and diagrid structure on 24° sloping ground.

3.1.1. Storey Displacement

Storey displacement is the overall displacement of ith storey with respect to ground. It is noticed that, the enhanced value of displacement is in conventional building i.e. 23.07 mm and Minimum value of displacement is observed in diagrid structure i.e. 9.988 mm as shown in Fig.4 (a) and (b). The storey displacement for the building model with diagrid has 43% less displacement compared to the model without diagrid system.

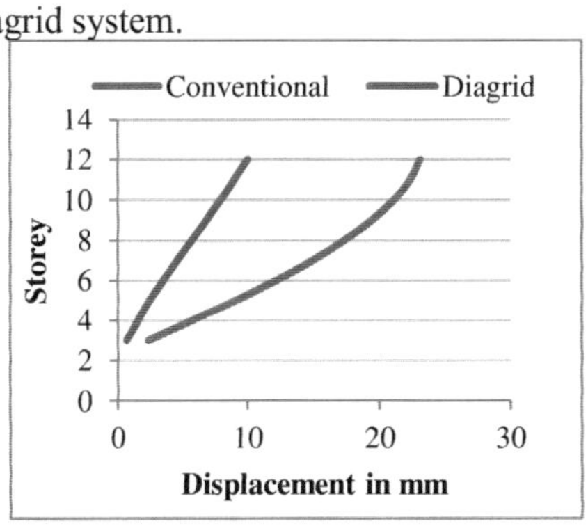

Fig. 4(a) Storey Displacement of Conventional building and Diagrid Building on slope.

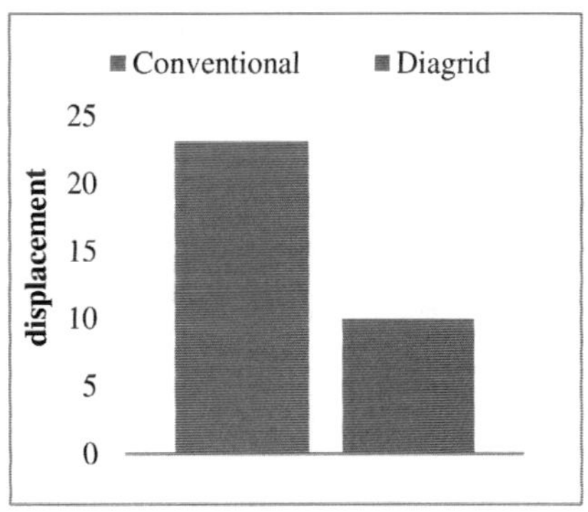

Fig. 4(b) Max Storey Displacement of Conventional building and Diagrid Building on slope.

3.1.2. Storey Drift

Storey drift is the relative displacement between floors above or below the storey considered. As per IS 1893:2016(Part-I) (clause 7.11.1) storey drift in any case shall not exceed 0.004 times of storey height. From the analysis, it is noticed that, the storey drift for diagrid structure is 35% less than the story drift of conventional structure as shown in Fig.5.(a) and (b).

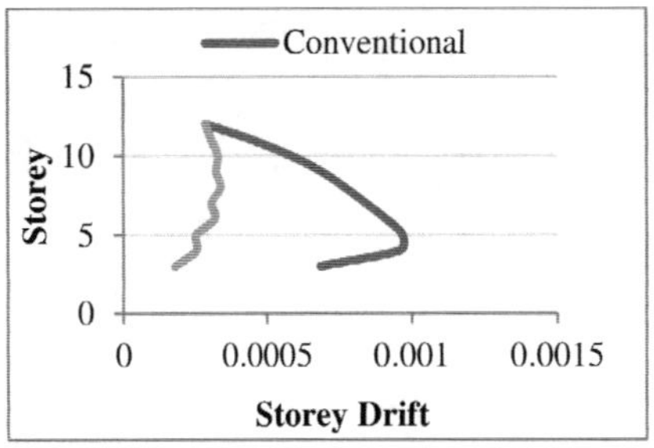

Fig. 5(a) Storey Drift for Conventional Building and Diagrid Building on slope

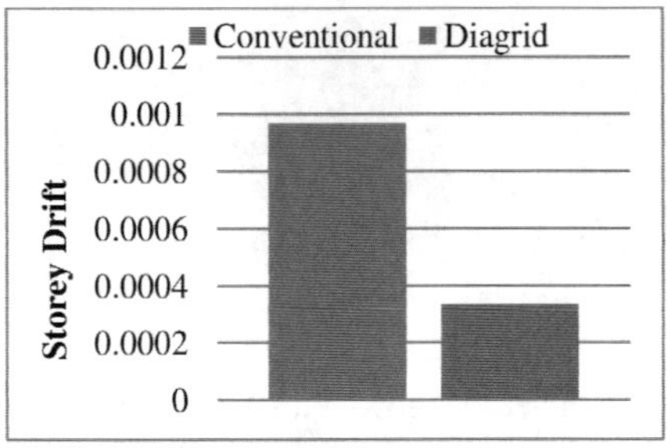

Fig. 5(b) Max Storey Drift for Conventional Building and Diagrid Building on slope

3.1.3. Storey Stiffness

Storey stiffness is the degree to which an object resists distortion in response to an usable force. Here it can be seen that the storey stiffness for diagrid structure has 35% more stiffness than conventional structure.

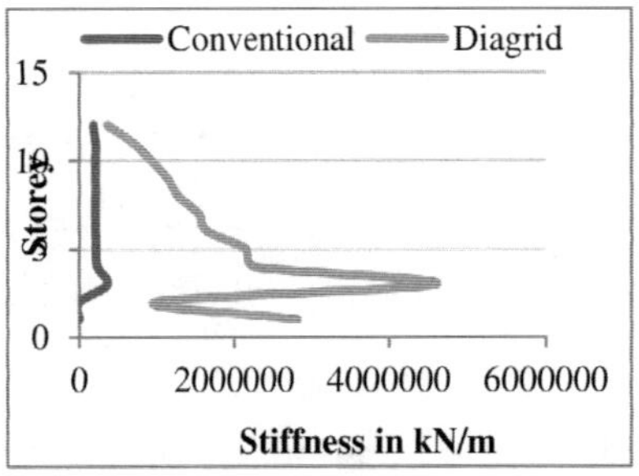

Fig. 6(a) Storey Stiffness for Conventional Building and Diagrid Building on slope

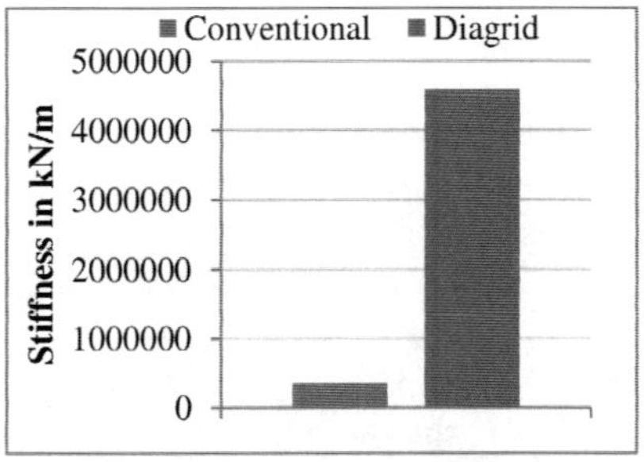

Fig. 6(b) Max Storey Stiffness for Conventional Building and Diagrid Building on slope

3.1.4. Base Shear

Base shear is an appraise of the highest await lateral force that will appear due to vibrating ground motion at the base of structure. It can be calculated by as IS 1893:2016(Part-I) clause 7.2.1. After the dynamic analysis from ETABS, evaluated Base shear are 853.64 and 2093.21 for conventional and Diagrid building respectively as shown in Fig.7. Also, it can be noticed that the base shear for the building model with diagrid system has 44% more base shear compared to conventional structure.

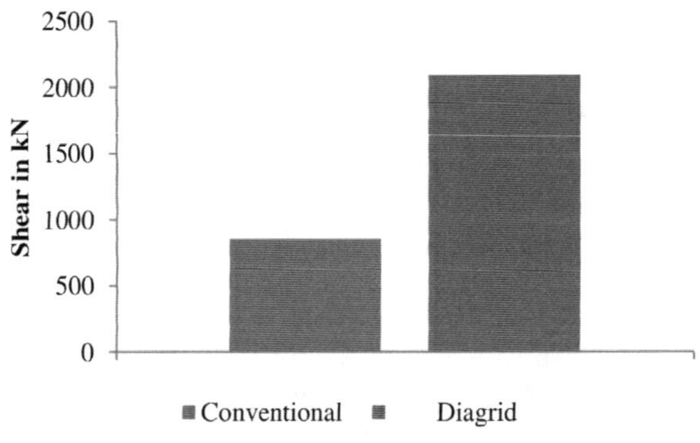

Fig.7 Variation of Shear force

3.2. Analysis of 48 storey structures with 15° and 25° slopes of ground for diagrid angles 53.47°, 60.94°, 69.67°, 74.47° and 79.5°.

A 48-storey diagrid building is considered for analysis and is comparing with different diagrid angles (54.47°,60.94°,69.97°,74.47° and 79.5°) on two different slopes of ground (15° and 25°) and the efficient angle on these two slopes of ground is estimated as follows.

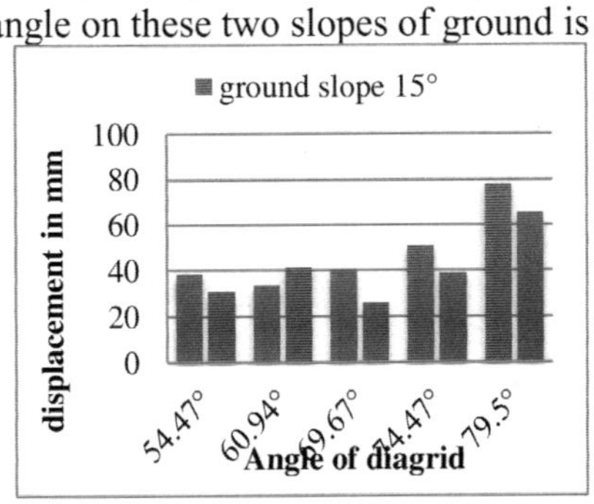

Fig. 8(a) Storey Displacement for Diagrid building on 15° and 25° slopes respectively in X-direction

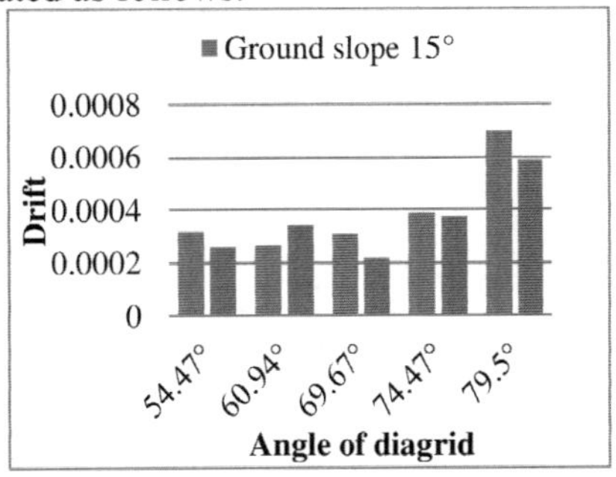

Fig. 8 (b) Storey drift for Diagrid building on 15° and 25° slopes respectively in X-direction

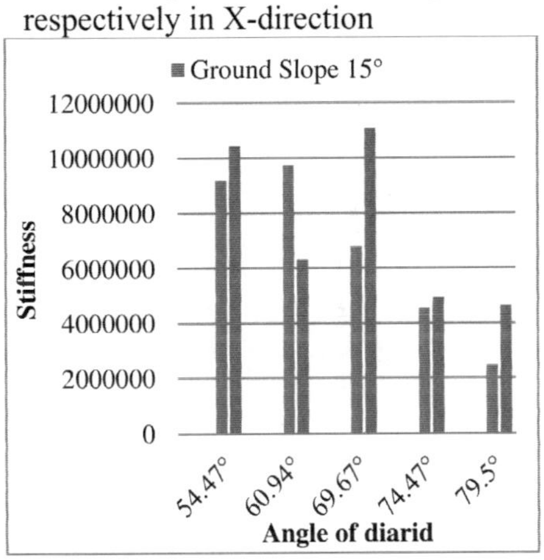

Fig. 8 (c) Storey Stiffness for Diagrid building on 15° and 25° slopes respectively in X-direction

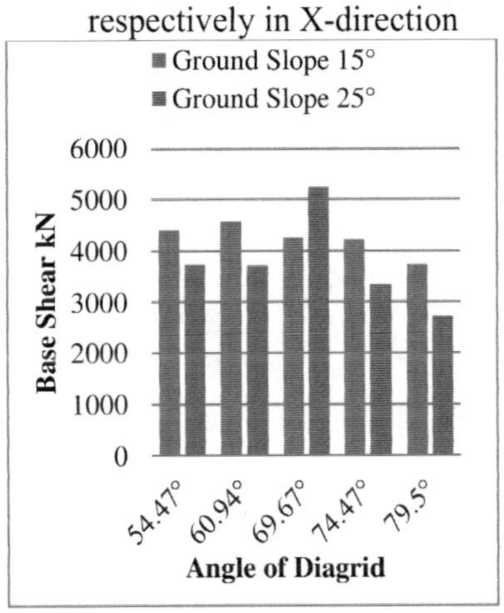

Fig. 8 (d) Storey Base Shear for Diagrid building on 15° and 25° slopes respectively in X-direction

It is observed from Fig. 8(a), the displacement of diagrid structure with diagrid angle 60.94° shows least displacement when the structure is built on 15° slope of ground but for 25° slope of ground, 69.67° diagrid angle shows least displacement compared to structures with other angles. From Fig.8(b) on 15° slope of ground the diagrid structure with diagrid angle 60.94° shows least storey drift, but for 25° slope of ground, 69.67° diagrid angle shows least displacement compared to structures with other angles. The displacement values in all the cases is (<0.004*h) and the drift values (<H/500) are within the limits.

It is also observed that from Fig. 8(c) that, the diagrid structure with diagrid angle 60.94° shows more stiffness when the structure is built on 15° slope of ground, but for 25° slope of ground, 69.67° diagrid angle shows more stiffness compared to structures with other angles. Also, from Fig. 8(d), the diagrid structure with diagrid angle 60.94° shows higher base shear when the structure is built on 15° slope of ground, but for 25° slope of ground, 69.67° diagrid angle shows greater base shear in comparison to structures with other angles.

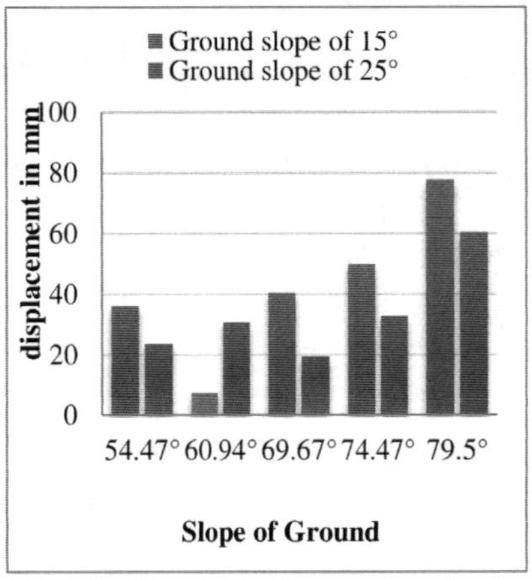

Fig. 9(a) Storey Displacement for Diagrid building on 15° and 25° slopes respectively in Y-direction

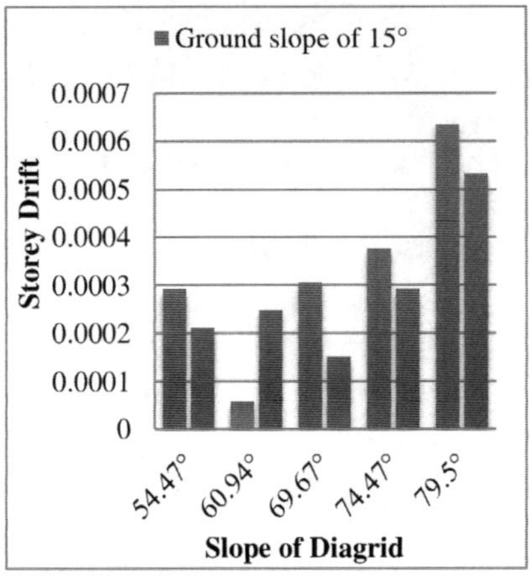

Fig. 9 (b) Storey Drift for Diagrid building on 15° and 25° slopes respectively in Y-direction

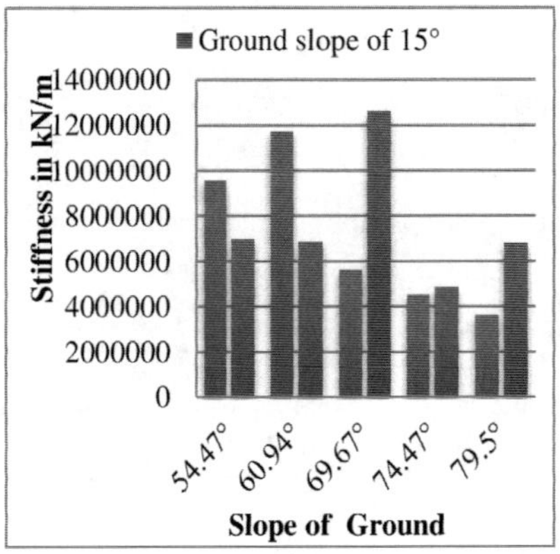

Fig. 9(c) Storey Stiffness for Diagrid building on 15° and 25° slopes respectively in Y-direction

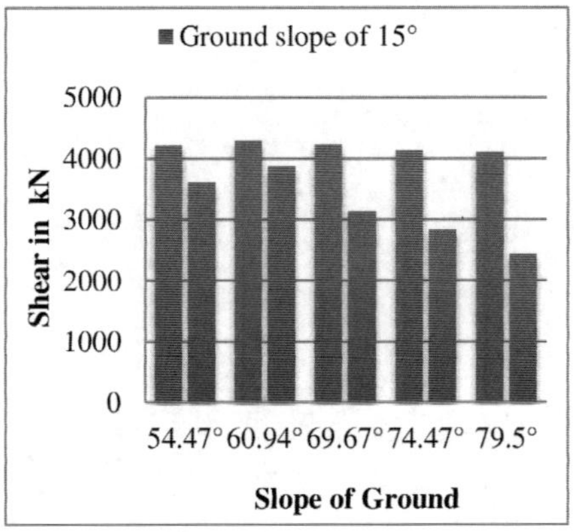

Fig. 9(d) Storey Base Shear for Diagrid building on 15° and 25° slopes respectively in Y-direction

It is observed from Fig. 9 (a), that the displacement of diagrid structure with diagrid angle 60.94° shows least displacement when the structure is built on 15° slope of ground but for 25° slope of ground, 69.67° diagrid angle shows least displacement compared to structures with other angles. From Fig. 9(b) on 15° slope of ground the diagrid structure with diagrid angle 60.94° shows least storey drift, but for 25° slope of ground, 69.67° diagrid angle shows least displacement compared to structures with other angles. It is also observed that from Fig. 9(c) that, the diagrid structure with diagrid angle 60.94° shows more stiffness when the structure is built on 15° slope of ground, but for 25° slope of ground, 69.67° diagrid angle shows more stiffness compared to structures with other angles. Also, from Fig. 9(d), the diagrid structure with diagrid angle 60.94° shows higher base shear when the structure is built on 15° slope of ground, but for25° slope of ground, 60.94° diagrid angle shows higher base shear compared to structures with other angles.

4. Conclusions

Based on the dynamic analysis of the different storey building on different sloping ground, the following are the conclusions from the analysis.

- Diagrid building showed lesser lateral displacement and storey drift compared to conventional building on sloped ground.
- Storey stiffness and base shear of diagrid is higher compared to conventional building.
- The seismic weight of diagrid structure is much lesser than conventional building. Materials used for diagrid are lightweight however provide high level of strength
- In 15° slope of ground, the storey displacement, storey drift is less for 60.94° (angle of diagrid) models. And for 25° slope of ground the storey displacement, storey drift is less for 69.67° (angle of diagrid) models.
- In 15° slope of ground the storey stiffness and base shear are high for 60.94° (angle of diagrid) models. And in In 25° slope of ground the storey stiffness and base shear are high for 60.94° (angle of diagrid) models.
- It can be conclude that for 15° slope of ground 60.94° angle of diagrid is the optimum angle and for 25° slope of ground 69.67° angle of diagrid is the optimum angle.
- Diagrid buildings show a good aesthetic appearance as compared to conventional building.
- Due to diagonal members on its periphery, Diagrid building shows good resistance to lateral loads.

References

[1] Moon, K.S., Connor, J.J. and Fernandez, J.E., 2007. Diagrid structural systems for tall buildings: characteristics and methodology for preliminary design. *The structural design of tall and special buildings*, *16*(2), pp.205-230.

[2] Moon, K., 2009. Design and construction of steel diagrid structures. *School of Architecture. Yale University. New Haven. USA*, pp.398-405.

[3] Kim, J. and Lee, Y.H., 2012. Seismic performance evaluation of diagrid system buildings. *The Structural design of tall and special buildings*, *21*(10), pp.736-749.

[4] Nguyen, B.K. and Altan, H., 2011, November. Strategies to Reduce Lateral Forces on High-rise Buildings that Use Diagrid Structural System. In *Proceedings of the 2011 World Congress on Engineering and Technology—IEEE Press* (pp. 795-798).

[5] Maqhareh, M.R., 2021. The evolutionary process of diagrid structure towards architectural, structural and sustainability concepts: reviewing case studies.J. Archit. Eng. Tech. 3, pp 1-11.

[6] Moon, K.S., 2012. Diagrid structures for complex-shaped Tall Buildings. In *Advanced Materials Research* (Vol. 450, pp. 1489-1492). Trans Tech Publications Ltd.

[7] Devaraja, R. and Megadi, M.R., 2014. Analysis of Curved Perimeter Diagrid Lateral System. *International Journal of Engineering Research and Technology, ISSN, 22780181*.

[8] Panchal, N.B., Patel, V.R. and Pandya, I.I., 2014. Optimum angle of diagrid structural system. *International Journal of Engineering and Technical Research*, *2*(6), pp.150-157.

[9] Malik, M.A., Azeem, M.A. and Nazim Raza, M., 2018. Effect of Infill Walls on Seismic Performance of RC Frames. *International Journal of Scientific Research and Innovation*, *5*(2), pp.23-26.

Building Information Modelling and Augmented Reality for building energy systems visualisation

Avani P[1] and Bhavana B[2]
1 PG Student REVA University, Bangalore-560064, Karnataka, India.
2 Assistant Professor, School of Civil Engineering, REVA University, Bengaluru – 560064, Karnataka, India. (avaniprem26@gmail.com; bhavana.b@reva.edu.in)

Abstract. Augmented reality is an important aspect of today's technological environment, as it aids in the development of technical abilities in students and engineers. Construction is often the industry with the least relation to digitalization, and we are currently dealing with challenges with real-view objects. Because some Elements are buried beneath the fake ceiling and walls, these components, which include ventilation, air conditioning, heating, and lighting, are now only partially visible in the building environment[9]. As per this article, Owners can track the project and use it for predictive maintenance, facility management, and future refurbishment using a mobile.

Keywords: SDK- Software Development Kit, AR- Augment Reality, VR- Virtual Reality, Buildings Energy systems Visualization.

1. Introduction

BIM (Building Information Modeling) is an intelligent model-based technology that unites (AEC) Architecture, Electrical, and Construction professionals, allowing them to more efficiently design, build, and run buildings and infrastructure. Property owners can use BIM for predictive maintenance, asset tracking, facilities management, and future redevelopment. These are mostly utilised in clash detection, model and scheduling [3]. Although augmented reality (AR) has been around for decades, it has seen exponential growth and progress in recent years, and it is becoming more popular. AR aids in the visualization of a project before it is completed. This technology has advanced the construction business to the point where a client sitting at home can view his engineers' and architect's drawings and virtually experience the interiors of a project. Heating, ventilation, air conditioning, and lighting systems are among the energy systems found in buildings, with space cooling, heating, lighting, and water heating accounting for over 60% of total energy use[9]. In the construction business, these provide numerous benefits, including the ability to examine building designs. Even though BIM gives information on energy systems, the possibilities for visualizing and analysing building energy system performance and behaviour using BIM and AR are still being explored.

1.1 Building Information Modeling (BIM).

Building information modeling is a process, not software. BIM is all about data; it contains a huge amount of information that may be utilized to better-building construction or design. BIM encompasses more than a two-dimensional or three-dimensional model. It has seven dimensions that help in calculations, sustainability, and upkeep. BIM is more than simply Revit or Navis; it's a collection of tools. BIM is all about the model's attributes. Architects enter design information, and MEP professionals enter data in the same database. Once all of the information is in, everyone can see the full project and make modifications at the same time.

1.2 Augmented Reality.

For the last few years, Augmented Reality (AR) has been a hot topic in the construction industry. With this technology, a customer sitting at home can look at his engineers' and architects' drawings and virtually experience the interiors of the project. It is also used for tracking, labour training, safety, time management, and so on [6]. People nowadays are pressed for time [2], and everyone's inadequacy is that they expect everything to be done swiftly and well, so augmented reality was formed. Augmented reality combines a real-world vision with a virtual one made by a computer through the use of an augmented view [4] [9]. Using the camera and a display on any device, such as smartphones, webcams on PCs, iPads, and so on, we may build augmented reality.

For the study, the following objectives were selected:

- To model G+3 Residential building using BIM
- To track the performance of the project
- To enhance the maintaining process
- To develop an app that introduces Augmented Reality in civil engineering, which further adds to its use in education and construction firms.

2. Methodology

2.1 General:

The technique used in this study was created using a variety of information sources and literature publications and is structured as follows. The design and development phases of this project's methodology are divided into two sections. This prepared report has been analysed and updated to match the scope and objectives of the project, with a proper conclusion. The complete project is integrated with Revit, Unity Editor, Vuforia Portal, and Android SDK. Together, they speed up the design process while enabling architects and designers to test out fresh concepts and spot potential design issues before they become issues. The client may envision so much more about the impending project by giving the proposal life. The study demonstrates the benefits of improved three-dimensional (3D) visualisation [1]. This project's model can be visualized with a smartphone or tablet. The process includes modeling, using several programs, and designing an AR. The majority of these technologies have been employed on building sites for progress monitoring and flaw detection as they have progressed in recent years [5].

2.2 Methodology:

- To plan a residential building according to international standards with help of BIM processes.
- Developing the app.
- Visualizing HVAC Hidden Behind the Walls
- Model complexity reduction and the other is linked with BIM information transfer to an AR system.
- How can the model be visualized appropriately without losing important information?

IOP Conf. Series: Materials Science and Engineering 1255 (2022) 012009 doi:10.1088/1757-899X/1255/1/012009

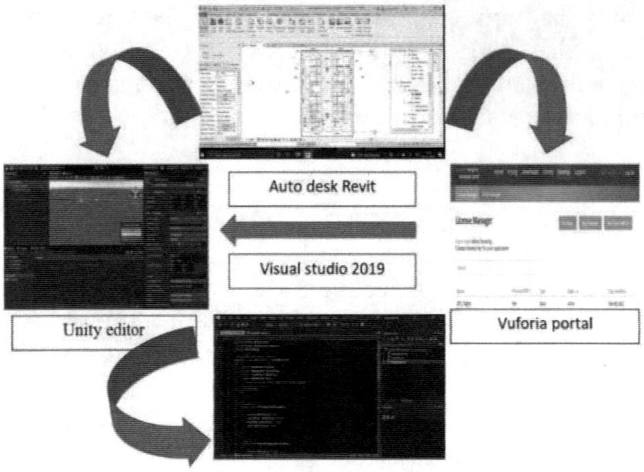

Figure 1. Methodology

3. Creation of a BIM Model

As illustrated in fig. 2, a Residential G+1 building was modeled for the experiment using Revit Architecture and Revit MEP (Mechanical Electrical and Plumbing). It was a 2700 square foot model with a 257 square meter built-up area. All of the basic materials were provided using Revit, and Revit MEP was utilized to design the ducts and plumbing. Grey ducts were installed, while the red pipe was installed.

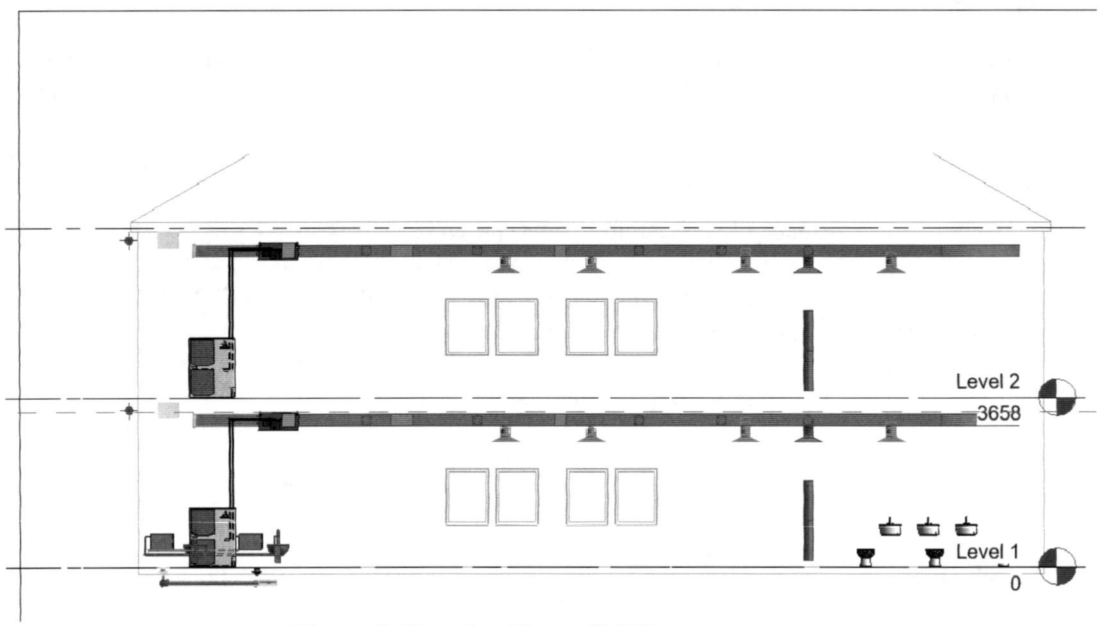

Figure 2. Elevation View of MEP components.

3.1 Working of AR:

The phone's front and rear cameras are used to generate augmented reality. A camera used to input both real-world and virtual items in real time, giving users access to a new kind of computing environment [7]. The ability to display distant items and information is intertwined with the task, and the user holds it up. The phone has evolved into a gateway to new world experiences and knowledge. That's because the same component that makes cell phones operate, such as accelerometers, is a computer vision mechanism that gathers visual input from the environment and sends it to the device's brain. Tracking for AR refers to the process of scanning, identifying, and interpreting ambient data. Outside engaging with the device using the g external camera to detect emotion and track position, there is two-way tracking going on inside out tracking and outside tracking outside interacting with the device using the g external camera to detect emotion and track position. Every application is intended to offer convincing virtual objects inside our tracking when the device can see an internal camera and sensors to detect motion and requests. Motion tracking SDK or software development kits, such as AR Core, Vuforia, and AR, which we utilize in this project, are one of the most crucial things to do.

3.2 Project Model:

Figure 3. Render Model.

.

4. Application Development

4.1. Augmented Reality Application Development for Android/IOS.

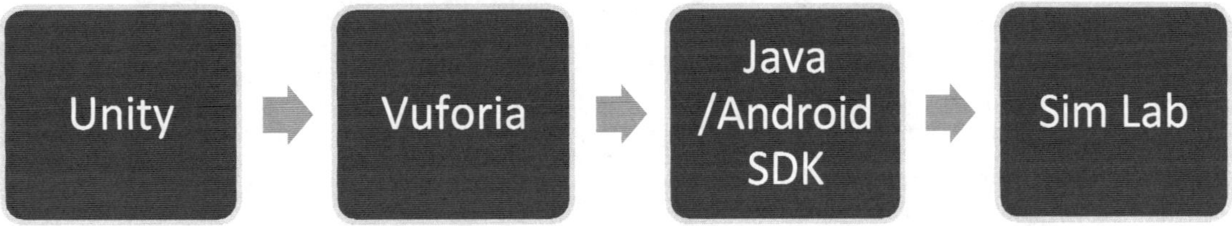

Figure 4. Application Development.

STEP 1:

Setting up 3D views in Revit, then exporting the needed files in FBX format after the model is complete. Using a free license, download Unity Hub and install the packages for Android or iOS, depending on the type of phone. Go to the Vuforia portal, download the Vuforia engine, load it into Unity, and add the required license.

STEP 2:

Create a new Unity project and delete the main camera by right-clicking on the hierarchy-Vuforia engine-AR camera. Import the 3D models into Unity using the import new asset command, then create a new folder under assets and add all of the button's images. Right-click-create-materials, rename the material, and in the inspector panel-shader-legacy shader-diffuse-insert the image-UI default Carry on with the procedure for the remaining buttons. Select AR camera, then open Vuforia configuration and enter the licensing information. Return to the Vuforia portal and choose target manager-add database-device-click on the database to build an AR model target and download the database into Unity. Create an image target site using the Vuforia engine, and the targeted picture will appear in the search results.

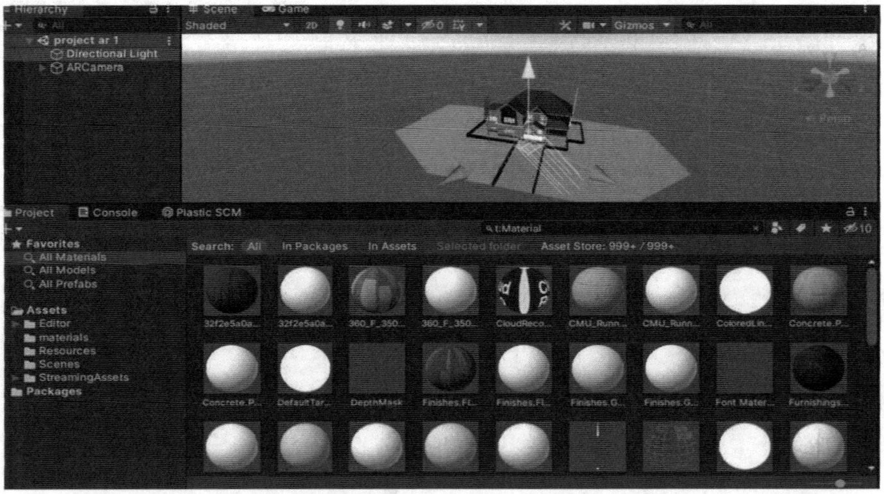

Figure 5. Unity Hub material inserting.

STEP 3:

Drag and drop all of the 3d models into the image target one by one, making sure to place them correctly using the scales. Right-click on hierarchy-UI-canvas to create the user interface; the frame's user interface appears on the scene. To make the buttons, right-click on canvas-UI-button and customize the button according to your needs.

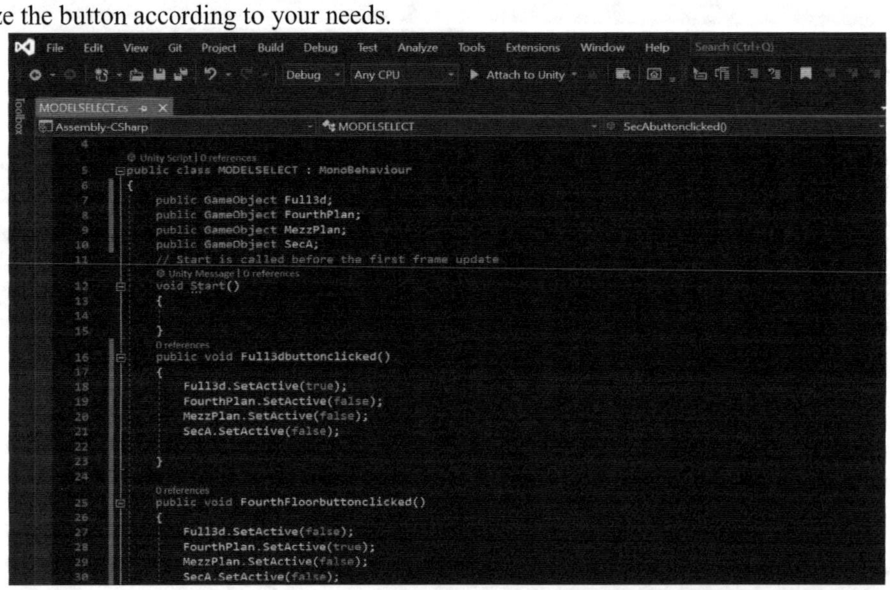

Figure 6. Visual studio.

STEP 4:
It requires scripting to make the interaction between the buttons, which is how your game is created. Create a C# script by right-clicking on assets and renaming it MODEL SELECT. When you click on a pattern, the 3D model associated with that button appears. Create an empty hierarchy by right-clicking on it and renaming it. Choose a model Attach the script to add components using a simple drag-and-drop method. The visual studio opens when you double-click the script. If you click the Model button, you may delete the unneeded function and type the script, save it, and return to Unity. Choose the variables that are visible. Return to the script and drag and drop all of the image targets into the variables one by one.

Figure 7. Augmented Reality.

STEP 5:
Make the necessary adjustments to the file-build settings-android-add scene-switch platform player settings-player. Click on build and run, make sure your Android is linked to your system, and then start the program from your mobile device.

Figure 8. Virtual Mode.

5. Discussion

5.1 How Augmented Reality Can Solve Construction Problems (MEP)

Mechanical, Electrical, and Plumbing (MEP) work is a critical component of the building industry. MEP focuses on environmentally friendly construction because it makes your building more energy-efficient. In general, in this industry, communication issues and outdated design are fairly widespread. As a result, in MEP building projects, data communication lines, lack of scheduling, confused revision, redrafting, inefficient resource use, and cost overruns are typical. Architects and engineers should be cautious, as even minor adjustments might quickly escalate into major problems. Models in 2D, 3D, and 4D are all distractions from the real world. No 2D model can portray how MEP looks in real-world conditions, either. Nonetheless, with augmented reality, we will be able to see exactly what it looks like. Assume you're in charge of a construction site. In most cases, every expert will have a precise plan of the work to be done, and the engineer now follows the plan to install a pipe at a certain location. When the HVAC crew arrives, they discover that the plans call for a duct to be put in the same spot, but the pipe is now in the way. At the end of the day, everything is a jumble. A professional BIM specialist can ensure that such issues do not develop; additionally, if all of this can be viewed using technology, Augmented Reality is born. Not just for AEC, but also for clients, everything has been easier using AR. Look for electricians, plumbers, HVAC technicians, and other professionals.

5.2 How Augmented Reality can change the MEP industry

Using Augmented Reality improves productivity compared to the traditional working methods. Reduce the time and resources it takes to translate between 2D plans in a 3D environment. A Huge number of conflicts can be resolved before the installation of equipment. Identifying the issues before they take place to reduce crashes and costly errors. This allows management teams to make intelligent business decisions that maximize workforce efficiency. Completing tasks faster and reducing the need for a second service call. In every field increasing customer satisfaction is all we need, MEP and HVAC systems require a lot of service calls, and AR improves the customer experience while reducing the cost.

6. Conclusion

In the actual world, BIM and AR have a big impact. Users can create unique digital experiences that mix the best of both the digital and physical worlds. In addition, the AEC business is embracing more AR technology to improve various stages of construction projects. AEC uses augmented reality to improve project control by visualizing construction progress immediately on-site. AR helps reduce Time, money, and quality as they are the most pressing issues in the building sector. Mechanical, electrical, and plumbing (MEP) play an important role in civil construction projects. MEP projects are characterized by increasing degrees of complexity as well as a wide range of transdisciplinary and interconnected operations involving numerous trades. As a result, timetable delays and cost overruns are common in MEP building projects. Hence by using AR the issues can be identified before they take place which reduces crashes and costly errors. This research concludes that by using BIM and AR we can communicate and collaborate more effectively and plan the project more safely. A digital transformation like AR and VR is now a requirement, and it will play an increasingly crucial role in a variety of areas, including raising productivity, assuring worker safety, and operational excellence, improving user experience and ensuring profit margins.

7. References

[1] Singh A R and Kumar V S 2018 Int. J. Prod. Lifecycle Manag *User behavior in AR-BIM-based site layout planning* vol. 11, no. 3pp. 221–244.

[2] Shaikh F S 2021 6th Int. Conf. Converg. Technol. I2CT 2021 *Augmented Reality Search to Improve Searching Using Augmented Reality* pp. 1–5.
[3] Zaher M, Greenwood D and Marzouk M 2018 Constr. Innov *Mobile augmented reality applications for construction projects* vol. 18, no. 2, pp. 152–166.
[4] Davila Delgado J M, Oyedele L, Demian P and Beach T 2020: Adv. Eng. Informatics *A research agenda for augmented and virtual reality in architecture, engineering and construction* vol. 45, no. May, p. 101122.
[5] Rankohi S and Waugh L 2013 *Rankohi2013.Pdf no.* 2005, pp. 1–18.
[6] Ahmed S 2019 Organ. Technol. Manag. Constr. an Int. J *A Review on Using Opportunities of Augmented Reality and Virtual Reality in Construction Project Management* vol. 11, no. 1, pp 1839–1852.
[7] Lee J Y, Kwon O S, Choi J S and Park C S 2012 Int. Conf. Inf. Sci. Appl. ICISA 2012 *A study on construction defect management using augmented reality technology.*
[8] Lin J 2021 *Augmented Reality Use in Construction - A Case Study of the Industry.*
[9] Dudhee V and Vukovic V 2021 *Integration of building information modelling and augmented reality for building energy systems visualisation.* pp. 83-89.

Use of Waste Plastic as Fine Aggregate Substitute in Concrete

Pushpa Lumina[1*], Pallavi G A[2], Bhavya C H[3], Vanishree S [4] ,Supriya C B[5]

1,Assistant Professor, School of Civil Engineering, REVA University, Bengaluru, Karnataka, India.2,3,4 Assistant Professor, Department of Civil Engineering, Sapthagiri college of Engineering, Bengaluru, Karnataka, India,5 Assistant Professor, Department of Civil Engineering , SSIT,Tumkur, Karnataka, India, (pushpa.lumina@reva.edu.in; pallaviagowda@gmail.com; Bhavych.1989@gmail.com; vanishiva.25@gmail.com; supriya.cb@gmail.com;)

Abstract. Solid waste disposal is a significant issue in our everyday lives as a result of the rapid population expansion in nations like India. One of the biggest environmental issues is the handling of solid waste. Plastic garbage is the waste kind that causes the majority of adverse environmental effects. Plastic must be recycled or reused because it cannot biodegrade. The study's goal is to examine the behaviour of concrete formed from recycled plastic materials as well as some of its physical characteristics. This study takes into account M20 cement concrete, which includes recovered plastic waste as a partial replacement for fine aggregate. Plastic was used to replace 5%, 10%, and 15% of the concrete in the casting of a concrete cube.

Keywords: Cement, Fine Aggregate, Coarse Aggregate, Plastic Waste

1. Introduction

Plastic is a class of high molecular weight organic molecule. Synthetic resin is another name for it. Resin is the primary component, while other additives are added to satisfy a variety of applications. There are two types of resins among them: natural resins and synthetic resins. At the same time, a crucial element, the performance of the resin influences the fundamental performance of the plastics. The fundamental characteristics of plastics are significantly influenced by additions (sometimes referred to as additives). It can lower manufacturing costs, change the performance of plastics, and improve the moulding and processing performance of plastic parts. The shape of plastic can be preserved at room temperature. For it to enter a, specific temperature and pressure requirements must be met. One of the least understood and most challenging aspects of plastics' negative environmental effects is their disposal. Ironically, one of plastic's most desired qualities—its longevity and resistance to decomposition—is also the cause of one of its biggest drawbacks in terms of disposal. The synthetic chemical bonds in plastic are extremely difficult for natural organisms to dissolve, which has a major impact on the durability of materials. Less than 10 percent of the total amount of plastic produced is efficiently recycled; the remainder is either burned in incinerators or dumped in landfills where it will rot for countless ages. 1. The production and burning of plastic will increase greenhouse gas emissions. 2. Polluting the soil 3. Water contamination 4. Marine debris 5. Trash stains 6. Dangerous to living things. For agencies in cities, domestic has grown to be a significant issue. Low-density polyethylene makes up the majority of the garbage plastic bags used for residential waste (LDPE). The initiative aims to develop cost-effective building materials and attempt to meet the societal requirement for safe and affordable waste management using an environmentally friendly way. This project is thought to be helpful in addressing issues related to the disposal of a significant volume of recycled plastic material, the reuse of plastics in the concrete industry, and is thought to be the most practical use.

2. Objectives:

1. Create a useful technique for using used plastics.

2. Use less natural resources, such clay, in the manufacture of concrete blocks.

3. To reduce soil and water deterioration and the risk of pollution, reduce and reuse the formation of waste plastics on land and in water.

4. To create materials that are affordable to the average person.

5. Less plastic in trash streams means less non-renewable resource use.

3. Materials and Methods

Materials:

CEMENT: A binder, or substance used in building to bind things together, cement sets, hardens, and clings to other materials. Rarely is cement used alone; rather, it is used to bind sand and gravel (aggregates). Ordinary Portland cement 53Grade was utilised since it met IS: 269-1969 specifications.
AGGREGATE (COARSE AGGREGATE): One of the most crucial elements of concrete is aggregate (coarse aggregate). Concrete gains bulk from coarse aggregate. The various aggregate properties can be impacted by the characteristics of the concrete. The coarse aggregate used in concrete serves as a filler and gives the mass of concrete its strength and homogeneity. The coarse aggregate utilised is 20 mm.
PLASTIC: Fine particles of plastic that can't be further broken down have been crushed. The primary component of these plastics is high-density polyethylene (HDPE). Plastic-contained concrete can be utilised for non-structural components that don't need a lot of compressive strength. Additionally, concrete can be used in numerous applications, such as pavements that need adequate water drainage and sports field flooring, thanks to the high permeability and minimal absorption caused by plastic.
Fine aggregate (M sand) For concrete buildings, manufactured sand (M-Sand) is a river sand additive. Crushing solid granite stone yields manufactured sand. Sand that has been crushed is fashioned like a cube with grounded edges, rinsed, and categorised as a building material. 4.75 mm is the reduced range of manufactured sand (M-Sand).

Methodology:

1. Before anything else, plastic waste needs to be gathered and separated from other trash.

2. If plastic garbage is damp or contains moisture, it should be dried. Dry plastic garbage is what we must use.

3. Next, we break down the used plastic into tiny pieces. Fine particles are created by crushing tiny particles.

4. We have the ability to thoroughly mix and create a mixture.

5. We used moulds to put the mixture into. Dry the mould off.

IOP Conf. Series: Materials Science and Engineering 1255 (2022) 012010 doi:10.1088/1757-899X/1255/1/012010

4. Results and Discussions

4.1 Compression Test

The results of the pressure test unequivocally demonstrate that as the proportion of plastic trash increases, the compressive strength value drops. After 28 days, the concrete cube with zero percent plastic waste displayed the maximum compressive strength value of 19.5 N/mm2. Following this are plastic trash with percentages of 5%, 10%, and 15%, measured as 18.88 N/mm2, 19.33 N/mm2, and 17.77 N/mm2, respectively. The values fall as the percentage of plastic garbage in waste increases.

Table 1 : compressive strength of different ratio of plastic waste at 7 days

Sl. No	Sample Strength (N/mm2)	Compressive load(N)	Cross Section Area A(mm2)	Compressive(P/A)
1	0	285*1000	22500	13.1
2	5	230*1000	22500	11.11
3	10	275*1000	22500	12.66
4	15	225*1000	22500	10

Table 2 compressive strength of different ratio of plastic waste at 14 days

Sl. No	Sample Strength (N/mm2)	Compressive load(N)	Cross Section Area A(mm2)	Compressive(P/A)
1	0	385*1000	22500	17.1
2	5	350*1000	22500	15.11
3	10	375*1000	22500	16.66
4	15	325*1000	22500	14.44

Table 3 compressive strength of different ratio of plastic waste at 28days

Sl. No	Sample Strength (N/mm2)	Compressive load(N)	Cross Section Area A(mm2)	Compressive(P/A)
1	0	440*1000	22500	19.5
2	5	425*1000	22500	18.88
3	10	435*1000	22500	19.33
4	15	400*1000	22500	17.77

4.2 Absorption Test

The plastic garbage cubes performed incredibly well in the water absorption test. The percentage of water absorption declines as the proportion of plastic garbage rises. The percentages fall between 14.46 to 9.52. 0% of the concrete cubes performed poorly since they absorbed 23.08% of the water. A block of high-quality concrete cannot absorb more than 20% water. Therefore, the inclusion of plastic trash in concrete blocks enhances their functionality.

Table 4 percentage of Absorption test

Sample	%Water absorbed
0%	23.08
5%	14.22
10%	11.43
15%	9.56

Fig1.Cubes were immersed in water

Fig2.pressure is applied until cracks appear

4.3 Efflorescence Test:

Alkali is dangerous when it is present in bricks because it absorbs moisture and leaves a grey or white film on the surface. This test is used to find out if bricks contain alkali. A brick is submerged in fresh water for 24 hours as part of this test. After that, it is taken out of the water and let to dry in the shade. If there is no whitish film on the surface, alkali has not been added to the brick. The presence of alkali is permitted if a whitish coating is discernible on roughly 10% of the brick surface. It's mild if it covers around 50% of the surface. If the amount of alkali is greater than 50%,

4.4 Efflorescence Treatment

1. One of the most common ways to get rid of efflorescence has been to clean the wall with high-pressure water for a long time. The salts may, however, seep back into the wall and eventually emerge as crystals since they are water soluble. As a result, removing the white residue from the walls with a stiff brush should be the first line of defence. The residual salt on the walls can then be removed using water pressure washing.
2. It is more challenging to remove efflorescence if the source is calcium carbonate or calcium sulphate. It is challenging to brush off with a brush since it adheres to the wall extremely stubbornly. In this case, an acid-based therapy is advised.

The efflorescence test also showed the concrete blocks' superior qualities. For all ratios, the concrete blocks' surfaces are either completely white or devoid of any trace of grey. Every concrete block was demonstrated to be free of soluble salts or alkalis. We can infer from this test that there was no alkali in this block.

Table 5 Results of Efflorescence Test

SAMPLE	NIL	SLIGHT	MODERTE	HEAVY	SERIOUS
0%	√				
5%	√				
10%	√				
15%	√				

Fig3. The concrete cubes were measured for 2.5cm

Fig 4. Distilled water was poured

5. Conclusion

As the percentage of waste plastics increases, the total strength drops. A concrete cube demonstrated the maximum compressive strength rating of 19.8 N/mm2 (0 percent plastic waste). Following this are plastic trash with percentages of 5%, 10%, and 15%, measured as 18.88 N/mm2, 19.33 N/mm2, and 17.77 N/mm2, respectively. The lessening of adhesion between the waste plastic and the cement paste is responsible for this. The cement paste and plastic particles don't seem to be connected very well. However, since the water absorption is less than 15% for all ratios, it appears that cement and plastic waste can be combined. Additionally, salts and bases are barely detectable in concrete block mixtures that include plastic waste. Waste plastic cube combinations' reduced compressive strength values suggest that they should only be used in circumstances requiring a minimum amount of workability. Applications for precast bricks, partition wall panels, channel linings, and other building materials frequently involve such circumstances. A suggestion for more research will place an emphasis on grinding the waste into a fine powder and mixing it in a ratio that results in the greatest packing density. The compressive strength and bond may be improved as a result, or a plasticizer may need to be added to the mixture to strengthen the bond between the plastic surface and the cement particle.

References

[1] Use of Waste Plastic in Concrete Mixture as Aggregate Replacement by Patil Pramod Sambhaji. Assistant Professor, Civil Engineering Department, R.C. Patel Institute of technology Shirpur District Dhule, Maharashtra, India

[2] A Review in sustainable plastic waste in concrete by Mohammed Fadhil Qasim, Dr. Zena K Abbas, Dr. Suhair kadhem abed. Journal of Engineering Number 12 Volume 27, December 2021

[3] Literature Review on Different Plastic Waste Materials Use in Concrete by Dr Muhammad Maqbool sadiq, Muhammad Rafique Khat. Assistant Professor, at NUST, Islamabad

[4] Innovative Techniques of Waste Plastic Used in Concrete Mixture by Pramod S Patil, J R Mali, Ganesh V Tapkire, H R Kumavat.International Journal of Research in Engineering and Technology eISSN:2319-1163|pISSN:2321-7308

[5] Use of waste plastic as fine aggregate substitute in concrete by Amalu.R. G, Azeef Ashraf, Muhammad Hashim, Rejith.K. U, Vijitha. International Journal of Scientific & Engineering Research, Volume 7, Issue 4, April2016..
[6] Utilization of Plastic waste for Making Plastic Bricks by R. S. Kognole, Kiran Shipkule, Manish Patil, Lokesh Patil. International Journal of Trend in Scientific Research and Development.
[7] Mechanical Properties of Concrete Using Recycled Plastic by Ahmed Abu Ishaiba.The Islamic University-Gaza Research and Postgraduate Affairs Faculty of Engineering Civil Engineering Department.
[8] Assessment of Mechanical and Durability Properties of Concrete Containing PET Waste by R K sharma, A Jain, T Gupta . Department of Civil Engineering, CTAE, MPUAT, Udaipur-313001, India.
[9] Comprehensive literature review on use of waste product in concrete B.V.Bahoria1 , Dr.D.K.Parbat2 , Dr.P.B.Naganaik3, Dr.U.P.Waghe4 Research Scholar, Civil Engg. Dept., YCCE, Nagpur, India
[10] Review on Utilization of Waste plastic in Civil Engineering Construction Projects by Aman Thakur, Agamjot SinghInternational Journal of Engineering Research & Technology (IJERT)
[11] Use of plastic waste in civil constructions and innovative decorative material (eco-friendly) By Mohammed Jalaluddin Jayaprakash Narayan College of Engineering, India Volume 3 Issue 5 – 2017
[12] Arora, A. and U.V. Dave, 2013. Utilization of E-Waste and Plastic Bottle Waste in Concrete. International Journal of Students Research in Technology & Management, 1 (4): 398-406.
[13] Hiremath, P.M., S. Shetty, P.G.N. Rai and T.B. Prathima, 2014. Utilization of Waste Plastic in Manufacturing of Plastic-Soil Bricks. International Journal of Technology Enhancements and Emerging Engineering Research, 2 (4): 102-107.
[14] Raju and R. Chauhan, 2014. An Experimental Study on Strength Behaviour of Cement Concrete with Use of Plastic Fibre. National Conference on Advances in Engineering and Technology, pp: 30-34.
[15] Raut, S.P., R.V. Ralegaonkar and S.A. Mandavgane, 2011. Development of Sustainable Construction Material Using Industrial and Agricultural Solid Waste: A Review of Waste-Create Bricks. Construction and Building Materials, 25 (10): 4037-4042.

Identification of Constraints in small construction projects

Sagar G[1], Raghunandan K[2] and Mehar Babu R[3]

1 M.Tech student, school of civil engineering, REVA University, Bengaluru-560064, India
[2,3]Assistant Professor, school of civil engineering, REVA University, Bengaluru-560064, India
(E-mail: sagargovindraj@gmail.com, raghunandan.koppad@reva.edu.in, meharbabu.ravula@reva.edu.in)

Abstract. Constraints in construction project limit the achievement of high performance in construction that can impact quality delivery, On-time completion and cost of the project. If constraints are understood well then better performance can be achieved by avoiding the constraints from impacting the performance of the scheduled project. In this current paper, the authors have done questionnaire survey from construction professionals and identified high probable constraints using Relative Important Index (RII) method. The Cost Overrun and Time Overrun caused due to the high probable constraints are segregated. Remedial measures are suggested to reduce or to eliminate the constraints caused in a construction project.

Keywords: Constraints, High probable, Measures, Relative Important Index.

1. Introduction

In every construction site there can be situations that we are not aware of the existence of the constraints, we tend to put more importance on the project goals. Even though constraints have been discussed in much of the management works, there is little detailed study on constraints in construction projects. However, we will not be aware of constraints which is going to occur in construction site or we focus more on goals of the project. The construction site includes multi party participation. In every multiparty working situation brings complication due to needs and constraints, which leads to conflicts and disputes, indirect or direct cost consequence to contractors and clients. The project team members should give equal importance to meet client expectations and to overcome the constraints occurred at construction site. For the constraints occurring at construction site identification of the potential constraints is important in construction project which will help to decrease the unnecessary wastage and loss of both money and time caused due the inadequate planning. If the constraints are managed well then high performance can be achieved in the project.

2. Literature Review

Vishal Y. Kolte, Himanshu Ahire (2020) [1] The main objective of the study is to identify the factors which causes delay in construction project and the effects caused by the factors on the project. To suggest the measures to overcome the delay. Questionnaire survey conducted through structured interview, the respondents were consultants, contractors, clients and stakeholders on a particular building project. There responses and views were gathered based on the factors that causes the main

reason for the delay of a project, and the results were documented and considered for designing any kind of organizational structure related to construction. The gap in this study was failed to do quantitative analysis, and the measures were given only for financial constraints and related delays.
Surbhi Rithe, Maithili Thakare (2020) [2], The main objective of the study is to suppress the constraints that causes time overrun, cost overrun, failure in quality of work. To do the research on these causes, find the reasons for its occurrence and finally to eliminate the causes, expose the importance of identification, suppression and elimination of a constraint in a project and to show its outcome and finally to improve its quality. The steps followed are identify, exploit, subordinate, elevate, repeat. Study finally concluded that having a good understanding and knowing the importance of identifying the constraint at the beginning stage and the documentation is important as it will be helpful for future projects to know causes for the delays and to elaborate step by step from the beginning of a project to its finish and all the delays encountered during the project and its effect on the profit of the project.
Anjay Kumar Mishra, Kailash Kumar Moktan (2019) [3], The main objective of the study is to identify the constraints specified related to a road construction project in its working environment and also to review about the possibility of Critical Chain Project Management (CCPM) approach to attain well planned scheduling. The steps followed to identify till eliminating the constraints are identify, exploit, subordinate, elevate and repeat. The study concluded that Sakosh -Tipling project encountered technical constraints majorly like mobilization, clearance of site, shoulder work, sub-base, base, joint survey and other general works. 35% of the work was delayed which led to 241 days of time overrun. The Bhimdhunga-Lamidanda project encountered legal constraints majorly which had 309 days of time overrun.
Desalegn Disasa Daba, Jayeshkumar Pitroda (2018) [4], The main objective of this study is to do a crucial literature review on the main causes of delay in construction project, a clear study on number of literature reviews done clearly explains the factors causing delay in a construction project and how it is related to the money. It also gives the information about a typical construction project, project time duration, project cost, required project quality and the elements involved in it such as cost, quality, quantity and time. It concludes that the main factors causing delay are consultant related delay factors and its causes, contractor related delay factors and its causes, owner related delay factors and its causes, external related delay factors and its causes and resources related factors and its causes are followed by labour related factors, construction materials and its related factors, finance related delays and its causes, equipment related causes, project related delays and causes.
Frank Ato Ghansah (2019) [5], The main objective of the study is to decrease the impact that is harmful or pathogenic on human. In the study the scientific reviewing technique is implemented. Research on the conference paper relevant to this study were considered by 56 papers with related to barriers to adopt technology. The barrier frequency was made use to choose the barrier which is reported most often which has a greater impact to increase the potency of the project and to reduce the waste generated, comfort, safety and accessibility for each building of an intelligent network. The gap is live wherever not given to overcome back the constraints. Quantitative analysis wasn't done.
Muhwezi L (2014) [6], Objective of the study is to evaluate the factors which is related for the delay in a construction project and also to study their effects caused on a construction project. The interview was conducted by sending the questionnaire survey for contractors, clients, consultants and engineers to know their point of views. The answers obtained help to rank the factors based on the ratings. The study concluded to have a control on the constraints that occurs in a construction project and its working environment, to achieve the better performance in a construction project. The gap measures were not given to overcome the constraints.
Gunduz M (2013) [7], The main objective of the study is to find out the factors causing delay at construction project, and to assess these factors by implementing (RII) Relative Importance Index method. Initially the factors were categorised to identify dealing factors, the factors were batched based on their priority with respect to its effect on delay. Prioritizing the factor which contributes the maximum for the delay. Identifying and eliminating the constraints or obstacles from the construction

activity, which in turn help to overcome the constraint or delays in the process of construction and it also improves the transparency as well as efficiency of a construction project. But the measures were not given to overcome the constraint.

Muhwezi L (2014) [8], Objective of the study is to evaluate the factors which is related for the delay in a construction project and also to study their effects caused on a construction project. The interview was conducted by sending the questionnaire survey for contractors, clients, consultants and engineers to know their point of views. The answers obtained help to rank the factors based on the ratings. The study concluded to have a control on the constraints that occurs in a construction project and its working environment, to achieve the better performance in a construction project. The measures were not given to overcome the constraints.

2.1. Background

Relative Importance Index method is adopted to determine the relative importance of the various causes of delays. In this case, rating ranges from 1 (equal important) to 9 (extremely important) and transformed to Relative Importance Index (RII) for each factors using the formula shown below:

- Relative Importance Index, RII=$\frac{\sum w}{A*N}$
- Relative importance index ranges b/w 0 to 1 i.e. ($0 \leq RII \leq 1$), higher value indicates more importance for that cause of delay.
- Where W = weighting given to each factor by the respondents which ranges from 1-9 where '1' is 'equally important' and '9' is 'extremely important',
- A = highest weight (i.e. 9 in this case), and
- N = total number of respondents (i.e., 43 in this case).

3. Data processing with questionnaire survey

Constraints are considered and questionnaire forms was set according to the factors that has been considered, where it consists of 30 questions based on both impact and probability and in total 50 sets of questionnaires were sent to the respondents to identify the most probable constraints and 43 number of respondents are received. Every question had a rating from 1 to 9 (1 is equally important and 9 is extremely important) and this survey form were forwarded to construction professionals like contractors, architects, engineers, and students who had experience of working at construction sites and the constraints that occurs at the construction site.

Table 1: The priority given to the ratings of the constraints

Importance of criteria	Rating	Importance of criteria	Rating
EQUAL IMPORTANCE	1	STRONG PLUS	6
WEAK OR SLIGHT	2	VERY STRONG	7
MODERATE IMPORTANCE	3	VERY, VERY STRONG	8
MODERATE PLUS	4	EXTREME IMPORTANCE	9
STRONG IMPORTANCE	5		

Table 2: Rankings obtained for various probability constraints factors by RII Method

Probability constraints factor	***ΣW***	**RII**	**Rank**
what is the probability of difficulties in obtaining work permits?	271	0.700258	1
what is the probability of failure in ownership of problems at site?	270	0.697674	2
What is the probability of improper allocation of money to the project?	266	0.687338	3
what is the probability of issues during land acquisition?	262	0.677002	4
what is the probability of constraints caused due to politics in construction project?	260	0.671834	5
what is the probability of failure in safety regulations at construction sight?	259	0.669250	6
what is the probability of difficulties in obtaining a loan from bank?	259	0.669250	7
what is the probability that can cause air, water, noise, dust pollutions due to construction activity	257	0.664082	8
what is the probability defective topography and soil strata at the site?	251	0.648578	9
what is the probability of poor planning and scheduling of the project?	249	0.643410	10
what is the probability of orthodox beliefs of people in construction project?	249	0.643410	11
what is the probability of inappropriate project cost estimation of the project?	248	0.640826	12
what is the probability of weather effect on execution activities of the project?	243	0.627906	13
what is the probability of not completing the project on time?	241	0.622739	14
what is the probability of imperfect drawings and details?	236	0.609819	15

Table 3: Rankings obtained for various Impact constraint factors by RII Method

Impact constraints factor	***ΣW***	RII	Rank
what is the impact caused due to inappropriate project cost estimation of the project?	291	0.75193	1
what is the impact caused by the issues during land acquisition?	285	0.73643	2
What is the impact caused due to improper allocation of money to the project?	283	0.73126	3
what is the impact caused by difficulties in obtaining work permits?	278	0.71834	4
what is the impact caused due to the failure in safety regulations at construction sight ?	274	0.70801	5
what is the impact caused due of failure in ownership of problems at site?	274	0.70801	6

what is the impact due to poor planning and scheduling of the project?	270	0.69767	7
what is the impact caused due to defective topography and soil strata at the site?	267	0.68992	8
what is the impact caused due to orthodox beliefs of people in construction project?	265	0.68475	9
what is the impact of constraints caused due to politics in construction project?	264	0.68217	10
what is the impact of weather on execution activities of the project?	264	0.68217	11
What is the impact due to imperfect drawings and details	260	0.67183	12
what is the impact caused due to air, water, noise, dust pollutions due to construction activity?	257	0.66408	13
what is the impact caused by not completing the project on time?	256	0.66149	14
what is the impact caused due to difficulties in obtaining a loan from bank?	255	0.65891	15

4. Results with conclusions

The factors associated with constraints are rated according to their ranking through the questionnaire survey by implementing Relative Importance Index (RII) method. The result is segregated into three categories, high, medium and low based on their rating given by the respondent.

4.1. Based on the Probability of constraints

4.1.1. *The factors under high rating are*

- LEGAL CONSTRAINTS - Difficulties in obtaining work permits.
- SOCIAL CONSTRAINTS- Failure in ownership of problems at site.
- ECONOMIC CONSTRAINTS- Improper allocation of money to the project.
- LEGAL CONSTRAINTS- Land acquisition.
- SOCIAL CONSTAINTS -Politics in construction project.

4.1.2. The factors under medium rating are

- LEGAL CONSTRAINTS- Failure in safety regulations at construction site.
- ECONOMIC CONSTRAINTS -Difficulties in obtaining a loan from bank.
- ENVIRONMENTEL CONSTRAINTS- Air, water, noise, dust pollutions due to construction activity.
- ENVIRONMENTEL CONSTRAINTS- Defective topography and soil strata at the site.
- SOCIAL CONSTAINTS- Poor planning and scheduling of the project.

4.1.3. The factors under low rating are

- TECHNICAL CONSTRAINT- Orthodox beliefs of people in construction project.
- ENVIRONMENTEL CONSTRAINTS- Inappropriate project cost estimation of the project.
- ECONOMIC CONSTRAINTS- Weather effect on execution activities of the project.
- TECHNICAL CONSTRAINTS -Not completing the project on time.
- TECHNICAL CONSTRAINTS -Imperfect drawings and details.

4.2. Based on Impact of constraints

4.2.1. The factors under high rating are

- TECHNICAL CONSTRAINTS- Inappropriate project cost estimation of the project.
- LEGAL CONSTRAINTS -Land acquisition.
- ECONOMIC CONSTRAINTS- Improper allocation of money to the project.

- LEGAL CONSTRAINTS -Difficulties in obtaining work permits.
- LEGAL CONSTRAINTS- Safety regulations at construction sight.

4.2.2. The factors under medium rating are
- SOCIAL CONSTAINTS -Failure in ownership of problems at site.
- TECHNICAL CONSTRAINTS -Poor planning and scheduling of the project.
- ENVIRONMENTEL CONSTRAINTS -Defective topography and soil strata at the site.
- SOCIAL CONSTAINTS -Orthodox beliefs of people in construction project.
- ENVIRONMENTEL CONSTRAINTS- Constraints caused due to politics in construction project.

4.2.3. The factors under low rating are
- TECHNICAL CONSTRAINTS -Weather on execution activities of the project.
- ECONOMIC CONSTRAINTS- Imperfect drawings and details.
- ENVIRONMENTEL CONSTRAINTS -Air, water, noise, dust pollutions due to construction activity.
- ECONOMIC CONSTRAINTS -Not completing the project on time.
- ECONOMIC CONSTRAINTS-Difficulties in obtaining a loan from bank.

Hence from the above observations it is concluded that all the constraints need to be identified and for project execution these constraints may be either avoided or suitable measure should be implemented as early as possible, especially legal, social and economic that are rated high in its probability of occurrence, where as technical, legal and economical constraints impact huge on the construction project according to the respondents of the questionnaire survey. The suitable measures are suggested to overcome constraints.

4.3. Remedial Measures

4.3.1. The following Measures are suggested to overcome the legal constraints
- Have a good communication with city permit department.
- Give a clear check on documents before the commencement of work.
- In case of any issues in the documents make sure to fix it immediately.
- Have a basic knowledge about local laws and regulations that regulates the construction project.
- Maintain a good bond with local municipality to have influence to overcome delays for any sanction related activities.
- Strictly follow the rules given by the government.
- Always have healthy communication with clients to make them understand the condition clearly that in turn creates a good understanding and avoids lots of issues.

4.3.2. The following Measures are suggested to overcome economic constraints
- Study the previous data of the same kind of project in past to know their expenses.
- After studying the previous data update the variables in case of any increase in the estimated cost.
- Never fix tight budgets.
- Do a research on the resource required, for better quality of labour's and material.
- Do a clear study on market rate for all the products and spot the most favourable deal.
- Always manage time, where time is directly linked with the cost.

4.3.3. The following measures are suggested to overcome environmental constraints
- Minimize the use of non-renewable energy sources such as fossil fuels.
- To minimise waste materials prefer buying building materials which are sustainable.
- Have a better plan to dispose the hazards waste.

- prefer 3R's (i.e.) reduce, recycle and reuse the materials to avoid wastage.
- Prefer making better plans for water conservation and water management
- Try to Style the project eco-friendly.

4.3.4. The following measures are suggested to overcome technical constraints

- Prefer updated technology for construction process to eliminate errors.
- opt skilled labours were ever possible.
- Choose good quality materials.
- Follow planning and building regulations strictly.
- Never prefer tight deadlines.
- Consider local climatic conditions before starting a work.
- Always have other alternative plans in case of failure of existing plan.
- Update each in every detail for clients in the process.
- Have a simple communication with labours for the better understand name.
- Regular inspection of work is required.

5. References

[1] Vishal Y. Kolte, Himanshu Ahire "Various Constraints for Delays in Construction Work and Suggesting Remedial Measures for the Same" International Journal of Research in Engineering, Science and Management Volume-3, Issue-1, January-2020
[2] Surbhi Rithe, Maithilee Thakare "theory of constraints in construction projects" International research journal of engineering and technology, volume-07 issue-12, December 2020
[3] Anjay Kumar Mishra, Kailash Kumar Moktan "identification of constraints in project schedule management" International research journal of granthaalayah volume-07 issue- 2, February 2019
[4] Frank Ato Ghansah, De-Graft Owusu-Manu Project management processes in the adoption of smart building technologies: "a systematic review of constraints Smart and Sustainable Built Environment'' emerald publishing volume-9 number -1, 2019
[5] Desalegn disasa daba Dr. jayeshkumar Pitroda "A critical literature review on main causes of delay in construction projects" International research journal of engineering and technology volume-05 issue-01, Jan 2018
[6] Muhwezi L., Acai J. and Otim G"An Assessment of the Factors Causing Delays on Building Construction Projects in Uganda" International Journal of Construction Engineering and Management, volume-03, issue-01 1, march 2014
[7] Gunduz M., Nielsen Y, and Ozdemir M. "Quantification of Delay Factors Using the Relative importance Index Method for Construction Projects in turkey" Journal of Management In Engineering, volume-29, issue-02, April 2013
[8] Rahman H. A, Takim R and Min W. S "Financial related causes contributing to project delays" Journal of Retail & Leisure Property, volume-08, issue- 03, August 2009

IOP Conf. Series: Materials Science and Engineering 1255 (2022) 012012 doi:10.1088/1757-899X/1255/1/012012

Compressive strength of recron fibred cement mortar by replacement of cement with alccofine

A S Deshpande[1], K N Nanjunda[2] and B Kavyashree[3]

[123]School of Civil Engineering, REVA University, Yelahanka, Bangalore, India (avin.deshpande@reva.edu.in ; nanjunda.kn@reva.edu.in ; kavyashreeb@reva.edu.in)

Abstract. In India, masonry structures are still practiced on major scale. Causalities majorly happens in masonry structures during earthquakes in earthquake prone zones due to less bond strength and non-structured designs. The main aim of afore mentioned paper is to find solution to increase this compressive strength, so that practice of masonry structures in rural area can made safe and economic with respect to strength. This study of bond strength can be carried out by analytic approach as well as experimental approach. In analytic approach, finite element method is used for the study, which requires large numbers of properties to model materials, which are not available in literature for the Indian scenario because literature focuses about results not properties. Hence this compressive strength study is carried out by experimental approach to give it more practical and meaningful direction. In mortar, arbitrarily dispersed fibers and alccofine 1203 are generally used for fortification to resist proliferation of crack and also helps to improve strength and ductility behavior which will help the structure to get less damage and to enhance the performance during earthquakes in earthquake prone zones. This will reduce the causalities. The use of recron 3s fibers is economic and alccofine 1203, easy to practice and effective approach as well. These experimental studies are started by making mortar cubes and ended up with making masonry prism and columns by using recron 3s fibers and alccofine 1203 with different mortar bed thicknesses. The studies on brick masonry prism suggests that use of recron 3s fiber and alccofine 1203 in mortar can increase the compressive strength to enhance the performance of the masonry structure. Use of 8 mm thick mortar bed gives more convincing when compressive strengths of masonry wallets are compared.

1. Introduction

Masonry comprises of singular Brick units and mortar head or bed joints. The Masonry is a Brick work build by regularly laying Brick binded together with mortar. The Brick blocks may be of marble, limestone, travertine, rock, cast stone, glass piece, solid square, and adobe. The Brick workmanship get influenced with the constituents' individual strengths. An Un-Reinforced Masonry may be non-load bearing or load bearing walls. For example, chimneys are URM structure made of brick, tiles, cinderblock, or other masonry that is unreinforced. Most of the structures in India, especially in rural areas are unreinforced structures which are vulnerable to earthquake forces. The bond strength between brick and mortar is not enough to hold for the lateral forces such as earthquake. Where these constructions are in locations with significant seismic risk is the key area of concern. Because URM structures are susceptible to collapse during an earthquake, the bond strength between mortar and brick should be strong enough to at least withstand the earthquake forces. One issue is that the majority of the mortar used to hold bricks together is insufficiently robust. Furthermore, brick components could "peel" away from the building and land on residents or bystanders outside. Particularly in areas that occasionally experience powerful earthquakes, there is grounds for concern. The strength of the bond between the mortar and the bricks is the key factor in masonry failure. It must be strong enough to withstand earthquake shocks for the building or unit to function as a safety precaution. Therefore, this research is being done to examine the bond's strength and determine whether it can be strengthened. Both an analytical and an experimental technique can be used to examine bond strength. For the study, which uses the finite element method analytically, a significant number of masonry properties are needed to model the material, but these qualities are not covered in the literature for the Indian context because the literature concentrates on results rather than properties. To offer this study a more relevant and useful orientation, an experimental approach was used to conduct it. Experimental studies were started by making mortar cube models and ended up with making masonry wall models for the study using recron 3s polyester fibres.

1.1. Need for the study

Masonry constructions are still used extensively in India. Due to its lack of structural support and lower compressive strength, masonry buildings in earthquake-prone areas are more frequently the cause of causalities. The above-mentioned project's primary goal is to find a way to boost compressive strength so that building masonry structures in rural areas can be done in a way that is both safe and cost-effective in terms of strength. To increase crack-resistant behaviour, this research, which has the same name, primarily focuses on the study of compressive strength utilising alcofine and recron 3s fibres in mortar. The compressive strength of the model was used to study the compressive strength. According to studies on brick masonry walls, adding alccofine and recorn 3s fibres to mortar can boost its compressive strength, improving the functionality of the masonry construction. Structures made of cement mortar are robust in compression but weak in tension. As a result, the structure will be less ductile, and cracks will spread more quickly. To fortify the cement mortar and avoid this issue, narrow steel fibre, mild steel fibre, alcofine, and recron fibres are typically employed. This will increase the brickwork's ductility and prevent fracture formation.

1.2. Objectives of the study are stated below,

- To develop suitable mix design for mortar with addition of alccofine & fibre.
- To determine optimum percentage of alccofine for maximum compressive strength.
- To find Alccofine's effect & also effect of fibre reinforcement on performance of masonry prism.

1.3. Mix design and materials used

For finding optimum dose of recron fibre and alccofine, fiber with different percentages of cement are added in the mortar fibre with constant percentage of cement are added in the mortor at 0.5%. For the study following mortar mix design is considered.
Mortar, alccofine & Recron Mix ratios used are given below:

- 1:6 (C: S) + 0% Recron 3s fibre + 0% alcofine
- 1:6 (C: S) + 0.5% Recron 3s fibre + 10% alccofine
- 1:6 (C: S) + 0.5% Recron 3s fibre + 12% alccofine
- 1:6 (C: S) + 0.5% Recron 3s fibre + 14% alccofine

1.3.1. Alccofine 1203. It is an microfine material with less calcium silicate chemical and is a porous material with self-healing properties. It can be used for the high-rise buildings, roads, bridges, ports etc., with more percentage over concrete industries. The main aim of the Alccofine is to reduce the water-cement ratio, chemical admixture and improve the compressive strength. Typical picture of Alccofine 1203 is shown in figure.1.

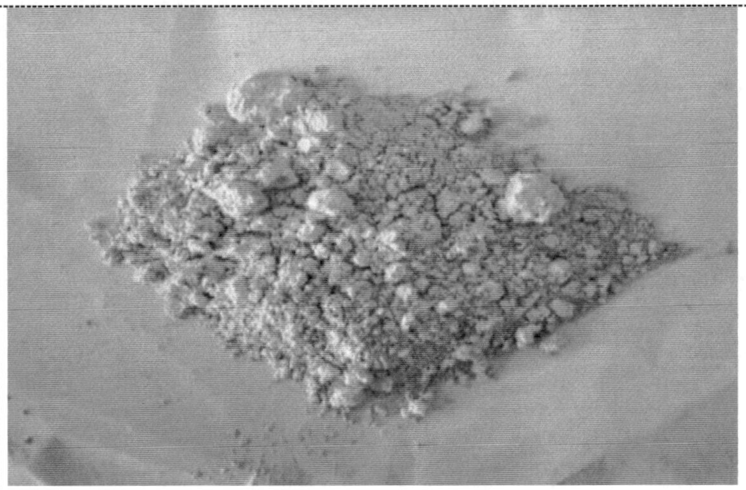

Figure 1. Alccofine-1203

1.3.2. Recron 3s fibre. It is a cutting-edge reinforcing material that is utilised to improve construction quality and boost strength in a number of applications, including automative batteries, paper, filtration fabrics, asbestos cement sheets, and precast items made of cement. Typical picture of Recron 3s fibre is shown in figure.2.

Figure 2. Recron 3s fibre

2. Methodology carried out

Figure 3 shows detailed methodology of the work carried out in this reserach.

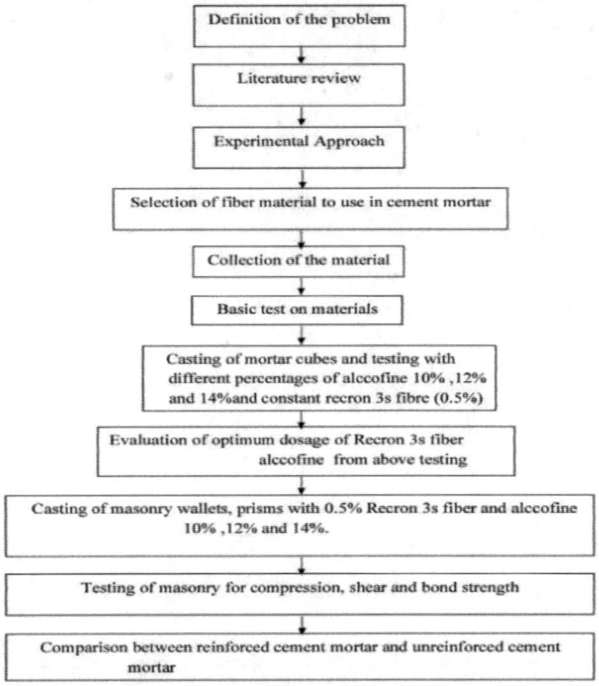

Figure 3. Detailed methodology of the experimental research work carried out

3. Results and Discussions

The results of the compressive testing of cement mortar cubes and the flow table test on cement mortar for various days are given.

3.1. Flow table test results

The results of the flow table test on cement mortar is shown in the table 1.

Table 1. Flow table test results on mortar

Trial No.	Designation	% of Fibre	% of alcofine	W/C Ratio	% of Flow
1				1.0	102
2	CM500	0.5	0	1.05	111
3				1.1	116
1				1.0	93
2	CM510	0.5	10	1.1	115
3				1.3	118
1				1.2	103
2	CM512	0.5	12	1.25	114
3				1.3	130
1				1.0	104
2	CM514	0.5	14	1.1	113
3				1.2	117

3.2. *Compressive strength test results*

Table 2, Table 3, and Table 4 show, correspondingly, the results of the compressive strength test on cement mortar cubes with recron fibre of 0.5 percent as constant and replacement of cement with varied percentages of alccofine for 3 days, 7 days, and 28 days.

Table 2. Compressive strength of cement mortar cubes at 3 days

Specimen size (LXBXH) in mm	Designation	Trial no	Comp. Strength (MPa)	Avg. Comp. Strength (MPa)
	CM500	1	1.48	
		2	1.52	1.53
		3	1.6	
	CM510	1	3.8	
		2	3.68	3.93
50x50x50		3	4.32	
	CM512	1	4.08	
		2	.4.0	4.28
		3	4.76	
	CM514	1	4.76	
		2	5.00	5.15
		3	5.68	

Table 3. Compressive strength of cement mortar cubes at 7 days

Specimen size (LXBXH) in mm	Designation	Trial no	Comp. Strength (MPa)	Avg.Comp. Strength (MPa)
	CM500	1	2	
		2	2.08	2.08
		3	2.16	
	CM510	1	4.48	
		2	4.40	4.49
50x50x50		3	4.60	
	CM512	1	4.88	
		2	5.04	5.08
		3	5.32	
	CM514	1	4.92	
		2	5.80	5.49
		3	5.76	

Table 4. Compressive strength of cement mortar cubes at 28 days

Specimen size (LXBXH) in mm	Designation	Trial no	Comp. Strength (MPa)	Avg.Comp. Strength (MPa)
	CM500	1	5.04	
		2	5.96	5.30
		3	5.92	
	CM510	1	9.81	
		2	9.67	9.80
50x50x50		3	9.92	
	CM512	1	9.96	
		2	10.02	10.08
		3	1027	
	CM514	1	10.01	
		2	11.36	10.88
		3	11.28	

Figure 4, 5 and 6 shows compressive strength of recron 3s fibred cement mortar cubes by replacement of cement by alccofine 1203 at 3, 7 and 28 days respectively in the form of bar charts.

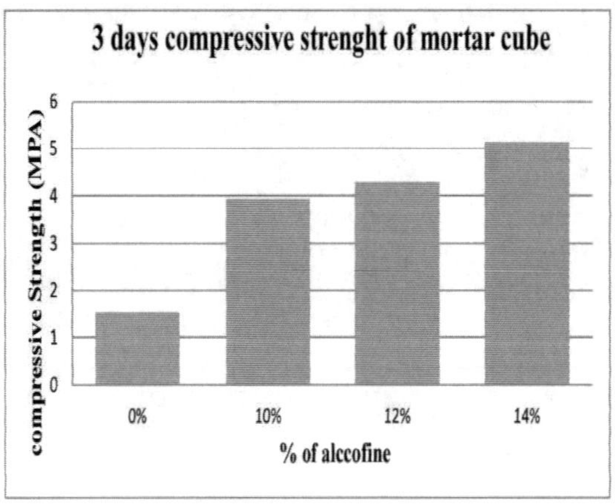

Figure 4. 3 days compressive strength of cement mortar cubes

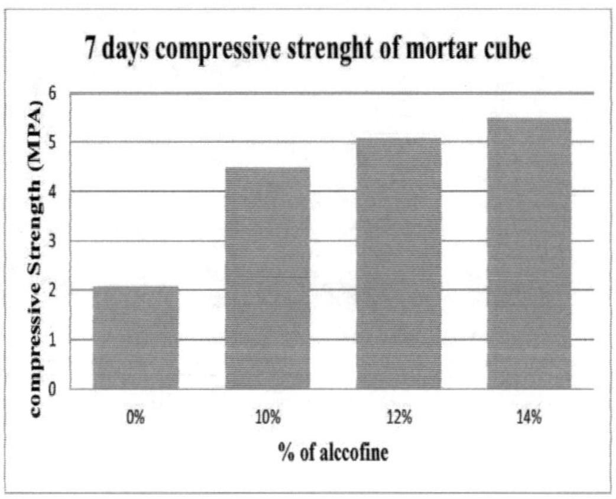

Figure 5. 7 days compressive strength of cement mortar cubes

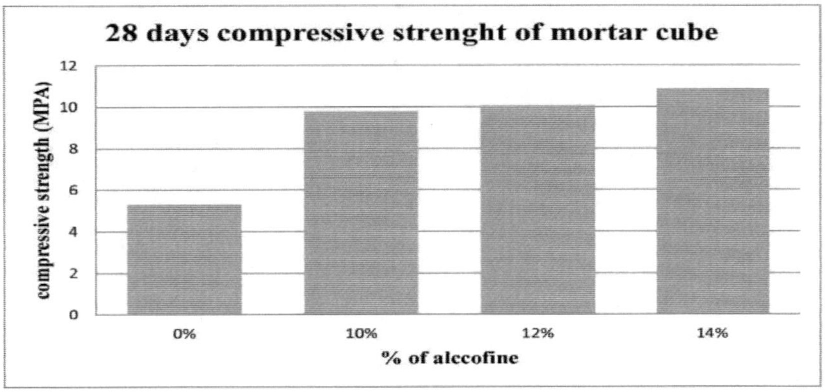

Figure 6. 28 days compressive strength of cement mortar cubes

4. Conclusion and discussions

- Compressive strength of cement mortar cubes for 0.5% recron 3s fibre as reinforcement is found to be maximum therefore the recron fibre value 0.5% is kept constant.
- Flow table values indicates that w/c ratio is above 1 for all the percentages of replacement of cement by alccofine in cement mortar.
- Percentage of alccofine 1203 variation is selected based on the previous literature studies i.e, experiments conducted in the past.
- Compressive strength of cement mortar cubes with recron fibre as 0.5% and cement replaced by alccofine of 14% is maximum at 3 days strength.
- Similarly Compressive strength of cement mortar cubes with recron fibre as 0.5% and cement replaced by alccofine of 14% is maximum at 7 and 28 days strength also.
- Further studies for future scope indicates that compressive strength tests can be done beyond 14% of alccofine replacement as replacement can be carried out in mortar cubes.

5. References

[1] B Kavya, K Rohith, Soniya "Experimental study on partial replacement of cement using Alccofine", ARPN Journal of Engineering and Applied Sciences vol. 12, No.9, May 2017.

[2] Malvika Gautam, Hemant Sood "Effect of Alccofine on strength characteristics of concrete with different grades". International Research Journal of Engineering and Technology, Vol.4, Issue 5, May 2017.

[3] R Rahul, G Manigandan, A Leema Rose "Study on strength characteristics of fly ash based Alccofine bricks". International Journal of Scientific Research and Education, Vol.5, Issue 9, September 2017.

[4] Manisha M Magdum, V V Karjinni "Influence of mineral admixture (Alccofine-1203) on the properties of hybrid fiber reinforced concrete". American Journal of Engineering Research, Vol.5, Issue 10, May 2016, pp 72-75.

[5] Siddharth P, Upadhyay, M A Jamnu "Effect on compressive strength of high-performance concrete incorporating Alccofine and fly ash", International Journal of Innovative Research and Development, ISSN 2278-0211, Vol.3, Issue 2, February 2014.

Acknowledgments
We would like to thank our organization REVA University for providing the necessary lab space so that the experimental investigation may be conducted. We would also like to thank the non technical persons i.e, laboratory instructors for their support in carrying out the works.

RISK ASSESSMENT IN BUILDING CONSTRUCTION PROJECTS

Rajesh[1], Vasanth Keshav[1]

[1]School of Civil Engineering, REVA University, Bangalore

ABSTRACT: Risks have a large impact on the budget, schedule, and a building project's quality. As the magnitude and complication of the project grows, the ability to control risk throughout the building process has become critical in preventing unplanned consequences. The high-quality procurement and the element of the corresponding agreement terms influence how risks are divided among project players to a great extent. As a result, project players must choose the best project procurement alternative. This dissertation's main objective is to not only assemble a table of building construction hazards, but also to identify the most important risks that can have a great influence on construction and mitigation methods. Assess and identify the most suitable countermeasure for the construction field in order to minimize hazards in future projects. The post-project appraisal method is applied in this dissertation to discover the list of hazards involved with the construction field. The questionnaire survey was conducted at several building projects, and the results were analysed once the survey was completed. The RRI method (Relative rank index) filters the response. Following this strategy, the AHP (Analytic Hierarchy Process) method is used to obtain more precise factors that affect project risk. The risks were discovered as a consequence of the analysis, and recommendations were made.

Keywords: Questionnaire Survey, Relative Rank Index (RRI), Analytical Hierarchy Process (AHP).

1. Introduction

Infrastructure improvements is one of the most significant things that may greatly strengthen the country's GDP by enhancing the businesses of various companies. Building projects are always inimitable, and risks can ascend from a wide range of sources. Many employees and management are associated in the building project; thus, the project's execution or completion may have an impact on their interests, either constructively or destructively. Members with different levels of expertise and experience may have different goals and interests. So, risk management has remained intensively researched in building project management in recent times. The Project Management Body of knowledge recommends nine knowledge fields, including risk management. Risk management is also a full and detailed system for recognising, assessing, and answering to risks to accomplish project objectives in the construction project management sector. Project completion in less time than expected is usually a difficult job for management since it typically requires several standard adjustments. Categorizing and analysing risks, and also improving Construction project management methods and resource efficiency, are all benefits of the Risk Management Process.

[1] employed principal component analysis and mean score analysis to rank the key risk factors based on their mean score produced from SPSS software with questionnaire survey data as input in this

study. This study identified and evaluated 6 main risk parameters that affect risk management in construction projects. The construction sector's risk parameters are listed by [2] by gathering information on building risks, their effects, and the educative procedures that may be performed to avert or reduce risk impacts. Risk analysis techniques were also investigated. [3] using SPSS, revealed that all of the samples (risk) ignored the null hypothesis. [4] concluded that almost all organizations experience a critical shortage of skilled labour. Project delays are caused by a lack of communication, a scarcity of materials, and design flaws. [5] summarised current research effort aimed at providing an experience and understanding method to risk management in building projects. [6] concluded that risk management must be involved preliminary in every construction project to get the most out of it. As a result, having a well-documented procedure that serves as a one-stop shop for any risks that may arise during the life cycle of the project is crucial.

[7] developed a framework for safety risk assessment and accident or injury anticipation costing, agenda that encourages consensus choice making among a group of decision makers while minimizing biased decision making using AHP. [8] showed that Lithuanian construction firms differ significantly from those in other countries in terms of risk management approaches. Combining probabilistic risk analysis approaches may enhance the risk management framework for building projects. [9] concluded that risk management in a well-executed project ensures a complete understanding of the service's goals, responsibilities, and features, as well as the projects prospective. It provides a data basis for quantitative data that is defined by size, with the objective of assisting judgements such as deciding between pricing and deploying items, or evaluating many possibilities. [10] focused on risk treatment and risk mitigation measures, especially in the context of construction project procurement. The authors identified strategies to projects, taking into account real-world scenarios and conditions, and brainstorming to reduce risk for their project participants. [11] offered a risk assessment approach and accompanying resources for subterranean building projects, including a risk assessment software system. [12] recognized eighteen major hazards connected with intercontinental construction projects in developing nations and classified into 3 levels of hierarchy (country, market, and project).

2. Methodology

A questionnaire was used as the research instrument in this study, and it was designed in such a way that the study's objectives were addressed in portions. The overall methodology of this study is identifying risk factors through literature and field observation, followed by their analysis, which is based primarily on data acquired through a questionnaire survey of various engineering and management people involved in building projects.

Among the different approaches for risk assessment and management, the Post Project Assessment method to identify the risks in the construction sector is considered. There were 45 questions and divided into ten categories. The questionnaire was measured using a 5-point Likert scale. Project Managers, Contractors, Engineers and purchasers from Bangalore were given these questionnaires. Microsoft form and Microsoft excel was used to collect and examine the data collected from the questionnaire survey. Relative rank index is used to rate criteria after gathering questionnaire surveys from various construction firms. AHP will be used to rank the criteria from top to bottom, and the most optimum project is selected using the same procedure.

The questionnaire survey is divided into three parts:

External risk: External risks are those that are out of the project management team's control.

- Political risk
- Economic risk
- Social risk
- Weather risk

Internal risk: Internal risk can be categorized according to who may be the source of risk occurrences, such as stakeholders, designers, and contractors.

- Resources risk
- Project members risk
- Construction spot risk;

- Documents and information risk

Project Risk: Project risks are those risks which occurred during the construction processes.

- Time risk
- Cost risk
- Work quality
- Constructions risk
- Technological risk.

Internal risk and project risk question have been considered. External risk has been not taken into consideration because they are out of project management team's control. The categories of risks considered in the study are listed in Table 1.

Table 1: Questionnaire Structure

Notation	Categories
	Time Risk
T1	No communication from top management to lower management
T2	Miscommunication from top management to lower management
T3	Improper Planning and Scheduling
T4	Inexperienced people involved in planning and scheduling
T5	Adoption of Wrong Methodology
T6	Inadequate communication for handling the changes in scheduling
	Cost Risk
C1	Improper Budgeting
C2	Delay in the delivery of materials
C3	Change in price of Construction Materials during the execution of the project
C4	Unexpected events delaying the execution of project (strike, elections, lockdown, etc.)
	Work Quality Risk
W1	Changes in the requirement of materials/plans during the execution of project
W2	Inexperienced workforce
W3	Demand for execution of work other than mentioned in the work-order
	Technological Risk
TE1	Errors in the design
TE2	Lack of required technology for the completion of project in the required time
TE3	Shortage of skilled manpower
TE4	Expectation of usage of better technology or equipment from the client during the execution of the project
	Resource Risk
R1	Shortage of Skilled manpower
R2	Shortage of required materials
R3	Change the during the execution of project due to changes in the schedule of rates
R4	Change in the during the execution of project due to changes in the requirements of the client
R5	Improper handling of materials during transportation
R6	Improper handling of materials at site
R7	Improper equipment allocation for the project
R8	Unexpected development near the site, causing delay in transportation of resources
	Project Member Risk
P1	Inexperienced personal for the execution of the project
P2	Insufficient Knowledge about work
P3	Non-cooperation between team members
P4	Lack of communication between workforce (staff and labours)
	Design Risk
D1	Design errors and omissions
D2	Improper quality check from the designer
D3	Lack of communication of design requirements to the team executing the project
D4	Failing to meet the prescribed standards, legal documents, and other conditions imposed by government agencies
D5	Stakeholder's request for changes during the execution of the project
D6	Design process takes longer than anticipated
	Stakeholders Risk
S1	Change the stakeholder during the execution of the project
S2	New stakeholder emerges and request for changes
	Construction Site Risk
C01	Unexpected site conditions
C02	Unpredicted Weather conditions

C03	Improper planning for the project without the consideration to site conditions
	Documents and information risk
D1	Contradiction in documents
D2	Loss of data
D3	Entering wrong information in the documents
D4	Improper maintenance of documents
D5	Improper communication of the required documents to the concerned authorities

3. Results and discussions

The survey findings and Relative Rank Index (RRI) values of various criteria as determined using the equation collected from engineers, project managers, quantity surveyors, structural designers, site engineers, and others are shown in Tables 5. Criteria are ranked according to their associated RRI values, with the greater the RRI value, the higher the rank, and vice versa. The factors with an RRI value (>0.73) are considered major risks in this investigation. The criterion whose RRI value is more than 0.73 and more are listed in Table 2.

Table 2: Major risk Criteria (RRI> 0.73)

Criteria	Condition	Ranking	RRI
C4	Unexpected events delaying the execution of project (strike, elections, lockdown, etc.)	1	0.803922
W2	Inexperienced workforce	2	0.780392
C3	Change in price of Construction Materials during the execution of the project	3	0.764706
R4	Change in the work order during the execution of project due to changes in the requirements of the client	4	0.760784
C02	Unpredicted Weather conditions	5	0.760784
D3	Lack of communication of design requirements to the team executing the project	6	0.752941
T2	Miscommunication from top management to lower management	7	0.74902
D2	Improper quality check from the designer	8	0.741176
C03	Improper planning for the project without the consideration to site conditions	9	0.741176
R1	Shortage of Skilled manpower	10	0.737255
R6	Improper handling of materials at site	11	0.730412

The pair-wise comparison matrix is prepared for each and every criteria by using Saaty's scale. To begin, two matrices will be created, one of which will be scaled from 1--9 and the other from 1-5. The Saaty's scale pair-wise comparison will be used to compare the first matrix. While making a pair-vice comparison, the Eigen-vector principle will take precedence. The reciprocals are allocated to the matrix after concurrently inputting the values in the matrix from 1-9. To obtain the Eigen values, divide each of the matrix's values by the column total in a pair-wise matrix.

Table 3. Pair-wise comparison matrix(1-5)

	C1	C2	C3	C4	C5	C6	C7	C8	C9	C10	C11
C1	1.000	0.330	0.330	1.000	0.500	2.000	0.330	0.500	0.330	0.250	1.000
C2	3.000	1.000	3.000	2.000	2.000	3.000	2.000	3.000	0.330	1.000	2.000
C3	3.000	0.330	1.000	3.000	4.000	0.500	0.330	0.500	0.250	2.000	2.000
C4	1.000	0.500	0.330	1.000	0.500	0.500	0.330	2.000	2.000	0.500	4.000
C5	2.000	0.500	0.250	2.000	1.000	0.330	0.500	0.500	0.250	0.330	0.500
C6	0.500	0.330	2.000	2.000	3.000	1.000	0.500	2.000	0.330	2.000	2.000
C7	3.000	0.500	3.000	3.000	2.000	2.000	1.000	3.000	0.330	0.250	3.000
C8	2.000	0.330	2.000	0.500	2.000	0.500	0.330	1.000	0.330	2.000	2.000
C9	3.000	3.000	4.000	0.500	4.000	3.000	3.000	3.000	1.000	2.000	3.000
C10	4.000	1.000	0.500	2.000	3.000	0.500	4.000	0.500	0.500	1.000	3.000
C11	1.000	0.500	0.500	0.250	2.000	0.500	0.330	0.500	0.330	0.330	1.000

Table 4. Pair-wise comparison matrix(1-9)

	C1	C2	C3	C4	C5	C6	C7	C8	C9	C10	C11
C1	1.000	0.660	0.330	1.000	0.200	4.000	0.660	0.250	0.140	0.500	1.000
C2	6.000	1.000	6.000	4.000	4.000	6.000	5.000	6.000	0.160	0.500	4.000
C3	6.000	0.160	1.000	6.000	8.000	0.250	0.140	0.200	0.110	0.160	5.000

IOP Conf. Series: Materials Science and Engineering 1255 (2022) 012013 doi:10.1088/1757-899X/1255/1/012013

C4	1.000	0.250	0.160	1.000	0.250	0.200	0.140	4.000	0.160	0.250	9.000
C5	5.000	0.250	0.120	4.000	1.000	0.160	0.250	0.200	0.120	0.140	0.200
C6	0.250	0.160	4.000	5.000	6.000	1.000	0.250	5.000	0.160	0.140	4.000
C7	6.000	0.200	7.000	7.000	44.000	4.000	1.000	7.000	0.160	0.120	6.000
C8	4.000	0.160	5.000	0.250	5.000	0.200	0.140	1.000	0.140	0.200	4.000
C9	7.000	6.000	9.000	6.000	8.000	6.000	6.000	7.000	1.000	5.000	7.000
C10	8.000	2.000	6.000	4.000	7.000	7.000	8.000	5.000	0.200	1.000	6.000
C11	1.000	0.250	0.200	0.110	5.000	0.250	0.160	0.250	0.140	0.160	1.000

Eigen Vectors, Priority Vectors and Sorted Weights are calculated by using the Table 3 and Table 4 matrix by Pairwise comparison matrix.

Table 5: Overall priority matrix

Criteria Representation	Criteria	RRI	Eigen Vectors	Priority Vectors	Sorted Weights
C1	C4	0.803922	0.047934	0.496185	10.35134
C2	W2	0.780392	0.129916	1.423978	10.96077
C3	C3	0.764706	0.084159	0.977894	11.61958
C4	R4	0.760784	0.079423	1.063094	13.38524
C5	C02	0.760784	0.043845	0.500985	11.42629
C6	D3	0.752941	0.090267	1.029652	11.40677
C7	T2	0.74902	0.116248	1.089812	9.374845
C8	D2	0.741176	0.0771	0.898458	11.65311
C9	C03	0.741176	0.174329	2.242495	12.86357
C10	R1	0.737255	0.109785	1.404442	12.79265
C11	R6	0.730412	0.046993	0.447371	9.519933

Now the Sorted Weights Column is filtered in increasing order, Top five questions are taken as top most priority. Hence we can say according to the survey C4, C9, C10, C8, and C3 are the major risk in the construction project. Next, Consistency Check is done. It is appropriate that the consistency ratio does not exceed 0.10. The judgement matrix is inconsistent if it is greater. To obtain a consistent matrix, judgement must be updated and refined.

Table 6: Consistency check

Consistency Check	
Size of matrix (n)	11
Principal Eigen Value(q)	11.39
Consistency index CI	
CI=	0.039582706
Average random consistency index for 11*11 matrix (RI)	1.53
Consistency ratio	
CR	0.02587105

4. Conclusions

The questionnaire survey was obtained from the project manager or engineers participating in construction projects at various locations. The Major risk in building projects was identified by analysing those questionnaires using AHP are:

- Change in the work order during the execution of project due to changes in the requirements of the client.
- Improper planning for the project without the consideration to site conditions.
- Shortage of Skilled manpower.
- Change in price of Construction Materials during the execution of the project.
- Unpredicted Weather conditions.

Based on the conclusions, the following recommendations are made:

- Presenting the suggestions for the shift's practicality based on understanding of the complete implications of the desired change or modifications. Changes will not be made in some circumstances because the perceived benefit is smaller than the expense. In other circumstances, there may be enough gain to cover the expense of undertaking the extra labour. In other circumstances, changes will be unavoidable and beyond control, regardless of their negative consequences, as a result of regulatory or compliance concerns, or internal issues such as organisational reshaping. It can affect both delivery and cost.

- Land factors (or site conditions) can influence the cost of construction as well as the layout or style of a home. The quantity of site preparation (or site works) necessary to establish the foundations for the home will have an influence on the cost of construction like, the kind of soil; the distribution of dirt over the block; How high is the water table, and are there any trees or rocks in the soil.
- Many organisations have begun giving training to their existing workers in order to adapt them to meet current shortages in order to compensate for a scarcity of trained people. This might include inhouse training when a skilled employee shares their valuable knowledge with others. It's a good approach for organizations with a few excellent resources to boost their potential without investing a lot of time or money.
- In order to prevent the effects of market increases in building material prices, contactor should have a well-developed material strategy in place ahead of time. To minimise cost overruns, conflicts, and inflation, effective planning and scheduling during the early phases of construction is critical, which includes the early acquisition of building materials within the planned cost and appropriate storage of building materials.
- Understanding the weather trends in the region and accounting for these predicted weather interruptions in a plan will help us avoid or at least lessen many of the delays. Weather-dependent activities can also be arranged for periods when better weather is forecast with proper planning. During Scheduling considering extra buffer time for unpredicted weather condition. We can also introduce mitigation measures to minimise the damage and get the project back up and running as soon as feasible.

References

[1] Srinivasan, N.P. and Rangaraj, A., 2020. Study on factors influencing risk management in construction projects. Management, 5(5.478), pp.18-259.

[2] Raj, M. and Wadsamudrakar, N.K., 2018. Risk management in construction project. International Journal of Engineering and Management Research (IJEMR), 8(3), pp.162-167.

[3] Vinodh K. R. and Therese A.V. 2017. Analysis of Various Types risks in construction projects. International Journal of Civil Engineering and Technology 8(7), pp. 401-409.

[4] Archana, V. and Francis, V.A. 2015. Project Assessment of Risks in Construction Industry. International Journal of Scientific & Engineering Research, 6(2),

[5] Serpella, A.F., Ferrada, X., Howard, R. and Rubio, L., 2014. Risk management in construction projects: a knowledge-based approach. Procedia-Social and Behavioral Sciences, 119, pp.653-662.

[6] Mahendra, P.A., Pitroda, J.R. and Bhavsar, J.J., 2013. A study of risk management techniques for construction projects in developing countries. International Journal of Innovative Technology and Exploring Engineering, 3(5), pp.139-142.

[7] Aminbakhsh, S., Gunduz, M. and Sonmez, R., 2013. Safety risk assessment using analytic hierarchy process (AHP) during planning and budgeting of construction projects. Journal of safety research, 46, pp.99-105.

[8] Banaitiene, N. and Banaitis, A., 2012. Risk management in construction projects. Risk Management–Current Issues and Challenges. In N. Banaitiene (Ed.), Risk Management–Current Issues and Challenges, pp.429-448.

[9] Schieg, M., 2006. Risk management in construction project management. Journal of Business Economics and Management, 7(2), pp.77-83.

[10] Khumpaisal, S., 2007. Risks in the construction project procurement process and the mitigation methods. Journal of Architectural/Planning Research and Studies (JARS), 5(2), pp.133-146.

[11] Choi, H.H., Cho, H.N. and Seo, J.W., 2004. Risk assessment methodology for underground construction projects. Journal of construction Engineering and management, 130(2), pp.258-272.

[12] Mustafa, M.A. and Al-Bahar, J.F., 1991. Project risk assessment using the analytic hierarchy process. IEEE transactions on engineering management, 38(1), pp.46-52.

IOP Conf. Series: Materials Science and Engineering 1255 (2022) 012014 doi:10.1088/1757-899X/1255/1/012014

A Case study of Urban Heat Island and Air Pollution Mechanism in Bengaluru Metropolitan city

Dr. T M Mohan Kumar[1], Suhas K B[2] and Dr. Siddegowda[3]

[1],Professor, School of civil Engineering, REVA University, Bangalore 560064.
[2],Research Scholar, Department of Civil Engineering, S J C I T, Chickballpur 562101
[3],Professor, Department of Civil Engineering, S J C I T, Chickballpur 562101

(Email: mohankumar.tm@reva.edu.in, suhasgowdakb@gmail.com , gowdas66@rediffmail.com)

Abstract. The industrialization in urban areas includes the economical increment, growth of GDPA, youth empowerment. The effects of industrialisation results in air pollution, sewage disposal, decline in green cover percentage. Urbanization or growth of city and Industrialization are embedded with one to another, without improving of Industrial sector, the growth of country cannot progress, but rapid industrialization and urbanization of metropolitan is cities of our country leads to generate huge environmental problems. This paper aims to determine and explore the relation with industrialization with Urban Heat Island (UHI). UHI is a phenomenon of accumulation of more temperature than rural areas. The metropolitan cities like Delhi, Mumbai, Ahmadabad, Kolkata, Bengaluru etc. These are the cities grown rapidly in present scenario. The mechanism of UHI, implications of it is in these metropolitan cities are overview of this research work. It includes collection, analysis of data related to UHI. Finally, it reveals the effect of UHI in different metropolitan cities. Increase in solid waste generation, air quality alteration, loss of aquatic life, air pollution, rise of skin diseases, asthma, respiratory problem etc. The objective of this paper is to understand the mechanism of the UHI process and its implications in different metropolitan cities by analyzing with parameter like raise in city temperature, thermal discomfort, pollution level, uneven rainfall, and increase in flood level.

Keywords: Urbanization, Urban Heat Island (UHI), Air pollution, Industrialization, metropolitan city.

1 Introduction

Industrialization:
Industrialization is beneficial to any country in its multi-dimensional growth. Thousands of Industries from other countries came to start up their Industry in this land. It positively benefitted to any country. In other face, it will generate so many complications to city environment like degradation of natural resources. Metropolitan is a large city with wide variety of people, culture, and character. Industrialization and

urbanization will further more influence on the habitat of the city. UHI discern most critical problem for the public in health issue.

1.1 Urbanization effect on Bangalore metropolitan city

Urbanization will made Rapid change in the morphology of the Bangalore. It also changes the land usage of Bangalore city. The growth of the city more than its existing limits leads to development of city in massive manner. The following table shows the land use of Bangalore city from last twenty years.

Table 1. Bangalore metropolitan city land usage.

Land Use/ Land Cover	1991	2011
Residential & Commercial	38.99	67.18
Layout	1.01	0.93
Industrial Expanse	1.93	4.97
Forest Cover	0.21	0.21
Barren Rocky	0.40	0.49
Water Bodies	3.16	3.04
Total Area in sq.km.	412	701

1.2 Urban Heat Island mechanism

It is a process of concentration of high heat waves in urban areas than rural areas. UHI is occurs due to Industrialization and urbanization. Industrialization is utilization of manpower to produce a new product. The highly developed country like Japan is more dependent on Industrialization itself. India is also coming same category in Industrialization, even though more dependent on agriculture. Because of urbanization the environment of the complete area of metropolitan city, get disturbed. In this contest, UHI comes under first one. The rise in temperature in urban areas not only discomfort to the people but also deteriorate the environment. Release of toxic

and harmful gases to the environment, uneven rainfall, heavy rainfalls, air pollution, and respiratory issues and so on.
Estimation shows that the overall population of the city reaches up to 3 billion. It is an indication that almost 40% of people are vulnerable to this dangerous effect. Anthropogenic heat is the one of main source of UHI, Vehicles releasing heat waves, temperature releasing from power plant, AC, refrigerator so on. Concrete buildings are also put up heat waves to the environment. Construction of various structures such as residential and commercial buildings requires massive volume of materials. All these are absorbent of heat radiations and reemits to the environment.

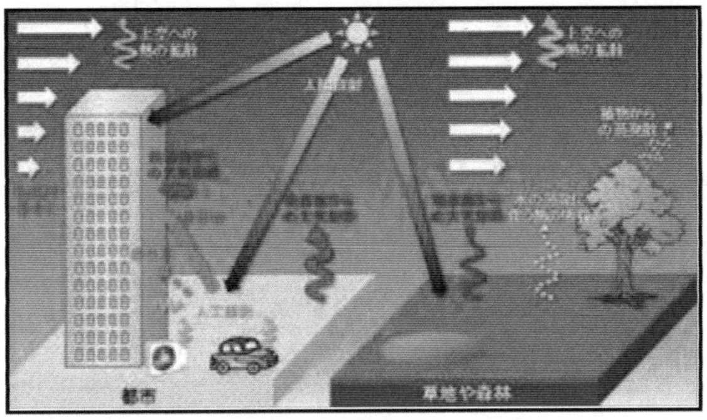

Fig. 1. Urban Heat Island Mechanism (source Google image)

1.3 Causes of UHI in metropolitan cities

The reasons or causes for UHI are not one factor oriented. It may differ from city to city; in general, few factors are considered as main factors causing UHI. Pavement surfaces made up of bitumen is one, which is dark in color, by its color itself absorb more radiation from the sun. The decline in the water resources like streams, lakes or ponds comes under this category to reduce heat accumulation. Drastic green cover loss is also added in to it. Encroachment of forest areas, parks, illegal construction, road widening are the few factors influencing green loss.

1.4 UHI effects in Metropolitan Cities

The UHI effect is have greater extent in metropolitan cities of our country. Here few effects are listed.

1. Rise in city temperature
2. Thermal discomfort
3. Increase in pollution level
4. Rise in electricity requirement
5. Uneven rainfall
6. Increase in flood level

IOP Conf. Series: Materials Science and Engineering 1255 (2022) 012014 doi:10.1088/1757-899X/1255/1/012014

2 Site Location and Methodology

Bangalore is the highly populated city in India. It is popular as silicon city, IT city, garden city etc. Bangalore is lies in southeast part of India. It are located at 12°58′44″N & 77°35′30″E and it covers an area of 741 km^2 (286 sq. m). The Bangalore Urban district has majority part of the city where as surrounded rural areas coming Bangalore Rural district. Generally, Bangalore is consisting flat area. The highest elevated point is located at the North West part of the city. There are no major rivers flowing in the city. Even though Arkavathi, south Pennar are the small river in Bangalore. Bangalore has other freshwater lakes like Hebbal Lake, Ulsoor Lake, Yediyur Lake and others. Bangalore consists of red laterite and red, fine loamy to clayey soils.

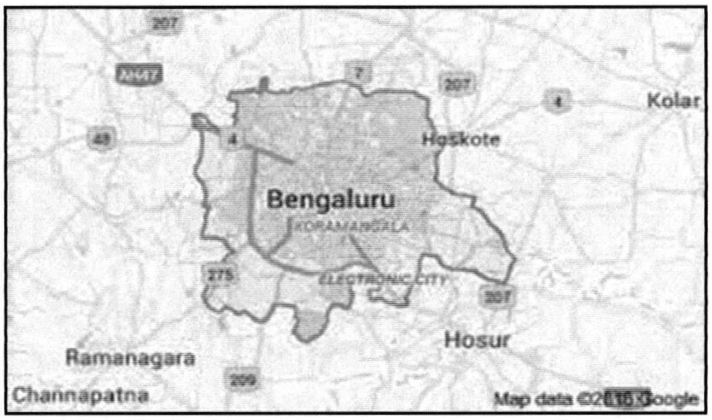

Fig. 2. Bangalore Google map

2.1 Methodology of the research work

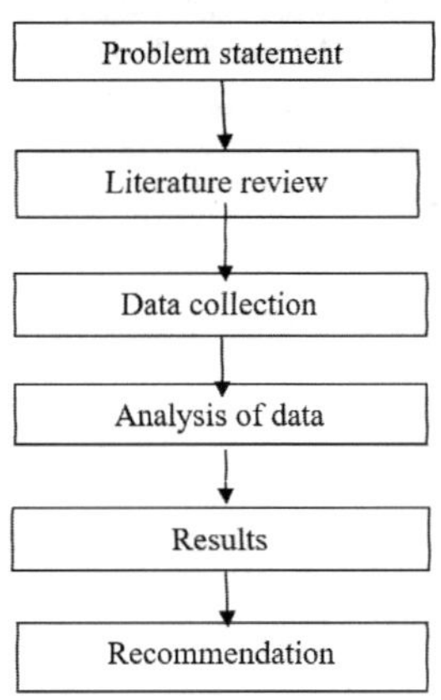

Step No 1: Problem statement

Urban heat island is the serious problem in Bangalore metropolitan city. The Industrialization will lead to increase in pollution level of Bangalore city. There are so many reasons for the increase in pollution level. Here the first step is to identify the severity of the problem. The UHI is the one of the big problem in Bangalore metropolitan city. Increase in pollution level is also making the problem more intensify.

Step No 2: Literature review

Literature review is also one of the major steps in this research. From the analysis of so many papers will give the idea on UHI as well as air pollution. The correct method to follow the work will gives by the literature analysis. In this approach, the present work has spotlight on so many research papers related on current topic.

Step No 3: Primary and secondary data collection

Once the literature review is completed, it is very important to focus on the collection of various data related to the research work. The data includes CO, NH3, NO2, ozone, CO2 etc. these data are carefully incorporated from available resources. The Indian pollution control board, metrological department, Disaster management, Indian population data and other related departments are approach to collect the needed data of research. Journals related to above topic from various publications are implemented in this research.

Step No 4: Analysis of data

After collection of the suitable data, it is necessary to analyse to get the expected results. The analysis wills mage by using available software's.

Step No 5: Results

The analysis shows the various factors involving air pollution as well as UHI effect. The various data have collected from different sources has to analyse to get the result. The result shows the reasons as well as remedial measures for UHI effect.

Step No 6: Recommendations

Once the result is over it is very important to focus on the recommendations. It is also play a vital role for further studies. A recommendation includes the remedial measures to minimize the effect of UHI.

3 Results and Discussion

Bangalore metropolitan city is an example for UHI effect. There are so many reasons influenced on the UHI effect. The Industrialization is also one of the reasons for rise in pollution level. Here the various data have collected and interpreted in the following tables and various graphs to show the various areas pollution level.

The present research work is mainly concentrated on Industrial area, residential area, and commercial areas along with residential area of Bangalore. Hebbal, Jayanagar, Railway stations are the examples of these areas.

In this result, the residential areas have shown min pollution level in the air quality whereas industrial areas have max level. Hebbal is residential and commercial areas so that it may slightly lower value of pollution.

More populated area as if Jayanagar has max level in pollution may be several reasons. The graphical representation of pollutants level understands these easily. The carbon monoxide, carbon dioxide, ozone level, PM level etc. is clearly shown in the following graphs.

Table 2. Bangalore metropolitan city land usage.

Parameter	Unit	Min Value	Max value	Average	Standard deviation	Number of observation
Benzene	ug/m^3	0.1	0.4	0	0	31
CO	ug/m^3	0.02	0.92	0	0	31
NH3	ug/m^3	6.5	10.4	8	1	31
NO2	ug/m^3	9.9	77.2	38	18	31
Ozone	ug/m^3	22.74	44.49	35	6	31
PM10	ug/m^3	93	141	110	14	31
PM2.5	ug/m^3	39	71	52	8	30

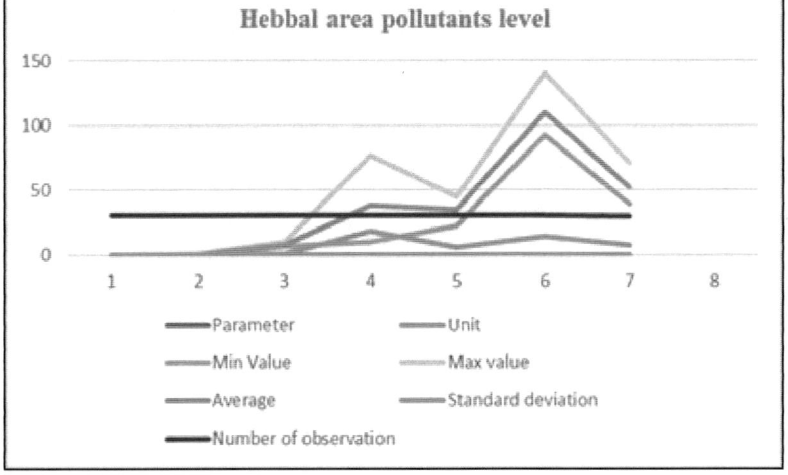

Fig. 3. Hebbal region pollutants level

Table 3. Jayanagar region pollutants level.

Parameter	Unit	Min Value	Max value	Average	Standard deviation	Number of observation
Benzene	ug/m^3	0.1	0.4	0	0	31
CO	ug/m^3	0.03	0.79	0	0	31
NH3	ug/m^3	13.3	15.9	14	1	31
NO2	ug/m^3	20.2	31	26	3	31
Ozone	ug/m^3	40.46	45.86	44	1	31
PM10	ug/m^3	77	127	102	16	31
PM2.5	ug/m^3	56	80	67	8	31

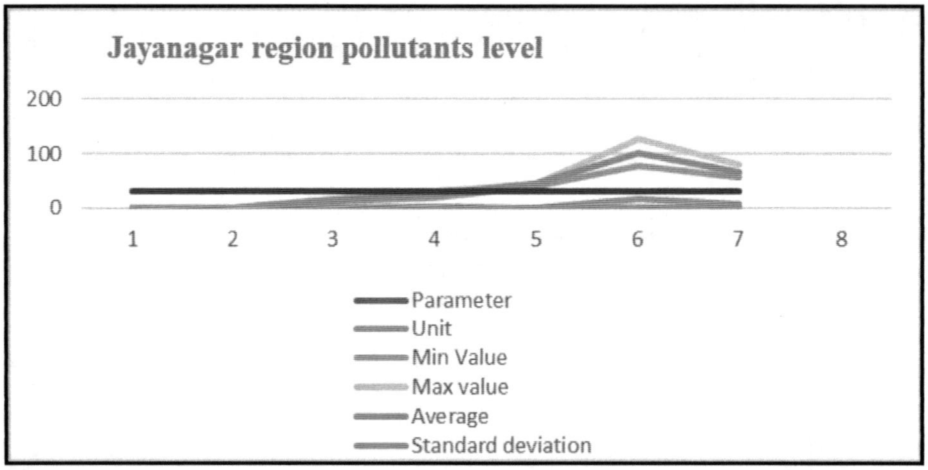

Fig. 4. Jayanagar region pollutants level

Table 4. Silk board region pollutants level.

Parameter	Unit	Min Value	Max value	Average	Standard deviation	Number of observation
Benzene	ug/m^3	0.1	0.3	0	0	31
CO	ug/m^3	0.19	0.76	0	0	31
NH3	ug/m^3	12.9	19.8	17	2	31
NO2	ug/m^3	22.3	54.1	35	7	31
Ozone	ug/m^3	31.48	56.15	44	5	31
PM10	ug/m^3	67	104	84	11	31
PM2.5	ug/m^3	45	69	57	7	30

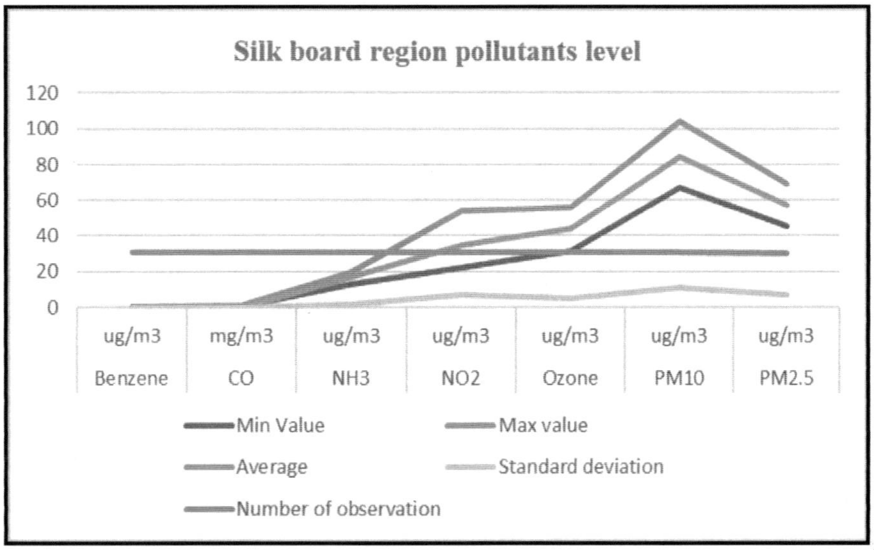

Fig. 5. Silk Board region pollutants level.

Table 5. Railway station region pollutants level.

Parameter	Unit	Min Value	Max value	Average	Standard deviation	Number of observation
Benzene	ug/m^3	0.5	5	2	1	31
CO	ug/m^3	0.03	1.08	1	0	31
NH3	ug/m^3	15.7	23.4	20	2	31
NO2	ug/m^3	13.4	32.8	20	4	31
Ozone	ug/m^3	42.7	94.1	74	13	31
PM10	ug/m^3	105	215	150	28	31

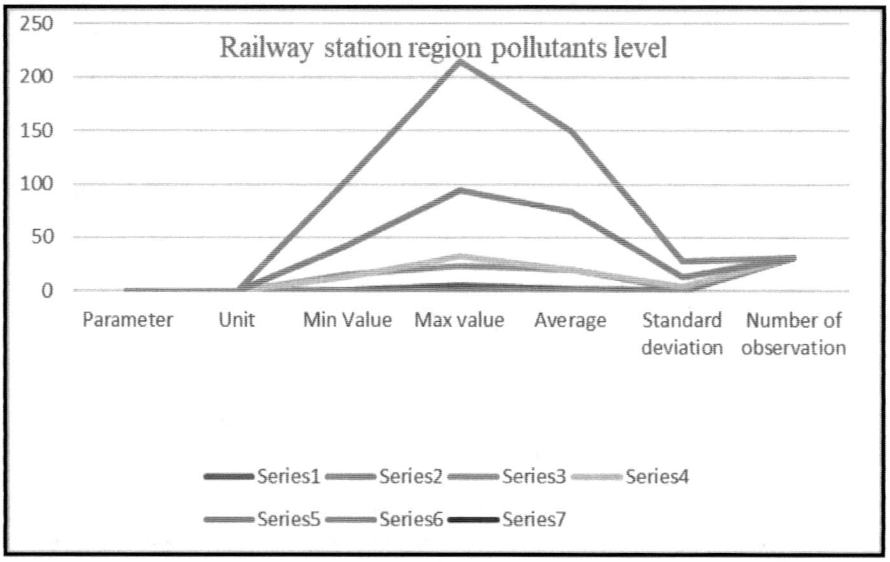

Fig. 6. Railway region pollutants level.

Table 6 . Bangalore region Temperature Information.

Area	Parameter	Unit	Min Value	Max value	Average	Standard deviation
Hebbal	AT	degree C	21.9	30	26	3
Jayanagar	AT	degree C	19	25.9	22	2
Silk board	AT	degree C	17.1	25.6	21	3
Railway station	AT	degree C	16.1	24.6	21	3

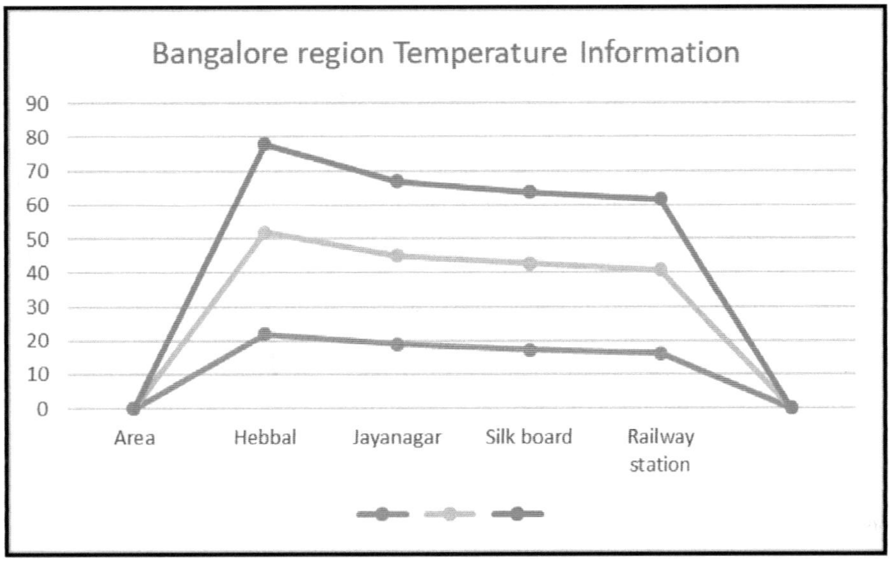

Fig. 7. Bangalore different regions temperature value.

References

1. Tayanc, Mete; Toros, Huseyin; Urbanization Effects on Regional Climate Change in the Case of Four Large Cities of Turkey, Climatic Change, 35, 1997, 501-524.
2. Ani, Che A I; Shahmohamadi, P; Sairi, A; Mohd Nor, M F I; Zain, M F M; Sura, M; Mitigating the Urban Heat Island Effect: Some points without altering Existing City Planning, European Journal of Scientific Research, 35 (2), 2009, 204-216.
3. Pomerantz, M; Akbari, Hashem; Pomerantz, B; The Effect of Pavements Temperatures on Air Temperatures in Large Cities, U.S. Environmental Protection Agency - Heat Island Group, Lawrence Berkeley National Laboratory, 2000, 1-12.
4. Hitchcock, David; Gero, Gary; Dietsch, Nikolaas; Cole, David; Reducing Urban Heat Islands: Compendium of Strategies, U. S. EPA Office of Atmospheric Programs, 2002, 1-52.
5. García Cueto, O R; Jaregui Ostos, E; Toudert, D; Detection of the Urban Heat Island in Mexicali B.C., Mexico and its Relationship with Land-Use, Atmosfera, 20 (2), 2007, 111-131.
6. Indian pollution control board values
7. Oke, T R; City Size and the Urban Heat Island, Atmospheric Environment, 7 (8), 1973, 769-779.
8. Akbari, H; Potentials of Urban Heat Island Mitigation, Proc. International Conference 'Passive and Low Energy Cooling for the Built Environment', Santorini, Greece 2005, 11-22.
9. The New Indian Express, Zeitgeist Supplementary Chennai, 8th July, 2009, 1.

10. Narayanan, Priya; Hanjagi, D. Ashok; Land Transformation: A Threat on Bangalore's Ecology-A Challenge for Sustainable Development, Theoretical and Empirical Researches in Urban Management, 4 (1), 2009, 38-47.
11. Ramachandra, T V; Kumar, Uttam; Wetlands of Greater Bangalore, India: Automatic Delineation through Pattern Classifiers, Electronic Green Journal, 1 (26), 2008, 1-22.
12. Krishna Iyer, Neelakantan; Kulkarni, Sumati; Raghavaswamy, V; Economy, Population and Urban Sprawl: A Comparative Study of Urban Agglomerations of Bangalore and Hyderabad, India using Remote Sensing and G.I.S. Techniques, Proc. Workshop on Urban Population, Development and Environment Dynamics in Developing Countries, 11-13 June, Nairobi, Kenya, 2007, 1-37.
13. Raghupathi, Usha P; City Development Plan for Bangalore under JNNURM, Proc. Regional Workshop, 24 January, Bangalore, 2006, 1-14.
14. The Hindu, Online edition of India's National Newspaper, 31st March, 2006.
15. Vagale, Uday Kumar; Public Space in Bengaluru: Present and Future Projections, Digital Libraries and Archives, Virginia Tech., 2004.
16. Ali, Sheik B; Tipu Sultan. National Book Trust, ISBN: 8123715862, 1995

GIS Integrated Ground Water & Sewage Analysis

Dr. M A Nagesh[1], Sissay Dechasa[2], Dr. G Veerappa Devaru[3] and Saksheshwari[4]

1 Professor, School of Civil Engineering, REVA University, Bengaluru – 560064, India.

2 Department of Water Resources Engineering, Adama Science and Technology University, Ethiopia.

3 Associate Professor, Department of Civil Engineering, SIT, Tumkur- 572103, India.

4 Assistant Professor, Sharanabasava university, Kalaburagi, India

(nagesh.ma@reva.edu.in ; dissiay@gmail.com, gvd@sit.ac.in, Sakshigec1@gmail.com)

Abstract. Groundwater is a natural resource for various human needs. It can be a major source for domestic and non-domestic use in any institution. The groundwater quality depends on percolation of rainwater and polluted water generated in an area, into the soils. The quality of ground water may not be potable in many cases. The freshwater demand could be met by demand management and reuse of treated sewage . The possible sewage reuse is emerging as potential source for demand management after essential treatment. In the paper an attempt is made to assess the quality of ground water based on BIS standards of water for domestic use. Also, the sewage water generated in the Sewage Treatment Plant (STP) is checked for its use for gardening use & possible ground water recharge. The results show that the ground water quality is good and satisfies drinking water standards. Also, the sewage from the treatment plant of REVA satisfies the quality standards for its use for irrigation or gardening purpose and possible ground water recharge in the campus. The GIS integration is also made for ground water & sewage water quality parameters

Keywords: Ground water, domestic use, sewage, recharge

1. Introduction

Groundwater is one of the sources of fresh water. Normally the ground water source can be used for drinking without much treatment. In 21st century our natural water resources have been used unconsciously which has led to it's over exploitation. The soil mantle through which water passes retain a large part of impurities and make the ground water free from the menace of pollution. Nowadays, emphasis is made to protect ground water and apply technology for its sustainable use. Also, the treated sewage can be used for the irrigation, gardening, fire protection, and toilet flushing in buildings. The treated sewage can also be used for ground water recharge for later use at a place.

The water scarcity problem of a campus can be overcome by ground water monitoring, rainwater harvesting & using treated sewage (Indu Nirupam Shukla et al.,2018, Dr sanjay D. Gaikwad, 2015, G.Elangovan et al.2019). Maintenance of the ground water level at higher levels can lead to effective use in summer season (S Rehan et al., 2014, S.S. Shinde, 2020, Meda Kalyan Kumar, 2015). The advanced technology of GIS & Remote sensing assists in achieving sustainability of water resources in the campus (Dr sanjay D. Gaikwad, A. S. Jasrotia, Abinash Majhi et al.). The treated sewage water can be used for the irrigation and also can be used for ground water recharge (G.Elangovan, et al., 2019).

Efforts are made to analyse the ground water quality status for domestic use & sewage water analysis of REVA campus, Bengaluru for use in campus for gardening and for recharging ground water. Laboratory analysis of ground water quality and sewage quality is done. The ground water is compared with the BIS standards for domestic use and treated sewage of campus is compared with the specifications for the use as irrigation water as per CPCB. The results are integrated using Geographic Information System (GIS) software.

2. Materials and Methods

An effort is made to check the quality status of ground water for domestic use and treated sewage water for gardening and/or recharging ground water in REVA campus, Bengaluru. Laboratory analysis is done for ground water and sewage for finding its quality. The Physico-chemical parameters analyzed in this study are, pH Total hardness as CaCO3, chlorides, Total Dissolved Solids, Alkalinity of both ground water and sewage. Along with this for sewage samples the analysis for Chemical Oxygen Demand (COD) & Biochemical Oxygen Demand (BOD) was done. The ground water is compared with the BIS standards for domestic use and treated sewage of campus is compared with the specifications for the use as irrigation water as per CPCB. The results are integrated using Geographic Information System (GIS) software.

The various ground water samples of campus at various locations(G1, G2, G3, G4, G5) were identified along with their GPS details. Also, at Sewage Treatment Plant (STP) entry sewage sample(S1) and exit sewage sample(S2) were taken for analysis. The samples were taken on 16.02.2022 and Table I gives the details of samples.

Table I. Location details of sewage water and ground water samples

Sample no.	Latitude in degrees	Longitude in degree	Description of location
S 1	13.115129	77.637072	STP Inflow
S 2	13.115597	77.637027	STP Outflow
G 1	13.114114	77.635195	Science Block
G 2	13.115701	77.634375	Sir. M V Block
G 3	13.115318	77.637230	Girls, G1 Block
G 4	13.117544	77.635785	Playground
G 5	13.115402	77.635121	Lawn

The REVA campus as in Google map is as in Fig. 1. The figure shows the campus study area being highlighted and the details of surrounding areas.

Fig 1: REVA Campus (Google Map)

3. Laboratory analysis

The laboratory analysis done for ground water and sewage sample of STP are given in Table II. The table also shows details of samples. The ground water was analysed for pH, TDS, calcium, magnesium, chloride, sulphate, Total hardness(TH) in terms of Calcium carbonate(CaCO3) & Total Alkalinity in terms of Calcium carbonate(CaCO3). Along with these parameters, for sewage samples additional analysis of Chemical Oxygen Demand (COD) and Biological Oxygen Demand (BOD) were undertaken.

Table II : laboratory analysis of sewage & ground water samples

Sl. No.	pH	TH mg/l	Ca mg/l	Mg mg/l	Cl mg/l	SO4 mg/l	TDS, mg/l	TA mg/l	COD, mg/l	BOD mg/l
S1	6.4	255	69.1	20	186	32	550	198	89.3	30
S2	7.01	219	56.5	19	157	25.3	550	166	19.4	5
G1	6.66	126	32.2	10.9	63.6	12.5	330	154	-	-
G2	6.6	7.8	1.96	0.71	13.6	10	330	128	-	-
G3	6.5	66.7	17.3	5.2	93.9	7.5	250	71.3	-	-
G4	6.5	82.4	24.2	5.2	95.8	9.8	312	194	-	-
G5	6.5	82.4	24.2	5.2	95.8	9.8	312	194	-	-

4. Geographic Information System (GIS) Integration

The GIS integrated image of REVA campus for Total Dissolved Solids (TDS) variation is shown in Fig.2. Excluding the sewage samples of high TDS located at the eastern side of the campus, it is observed that the ground water sample TDS value is high towards science block which is on the southern portion of campus.

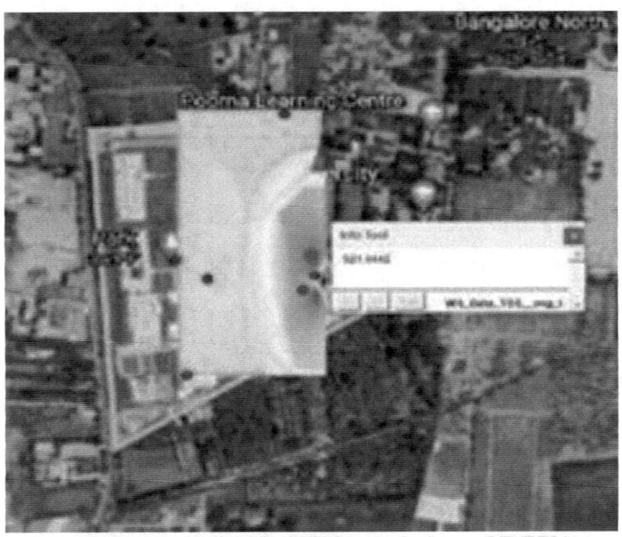

Fig 2: Total Dissolved Solids (TDS) variation of REVA campus

The analysed data is integrated with MapInfo GIS software using the data of ground water samples and sewage samples as in in table 2. The integration is done by using a google map of REVA campus along with xl sheet of data of samples analyzed. The GIS integrated image of REVA campus along with Information tool of a sample point is shown in Fig .3

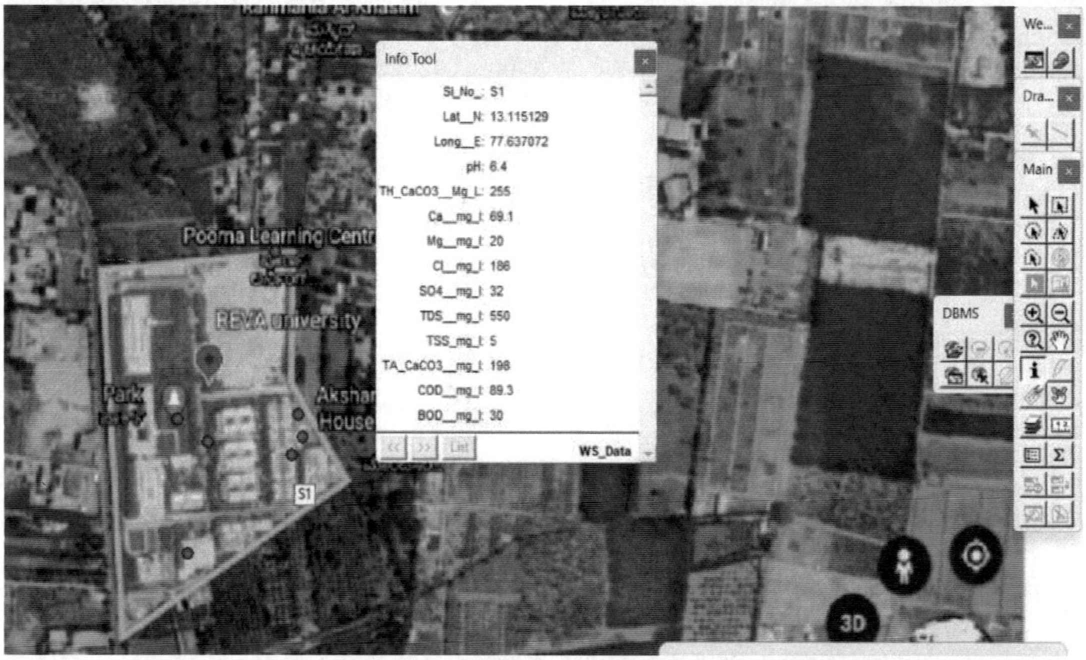

Fig 3: GIS integration of data with Information Tool for Water Quality

5. Results of samples analysed

The various parameters analysed for sewage samples are shown in fig. 4 & ground water samples along with sewage samples in Fig. 5, for various parameters like pH, Total Dissolved Solids (TDS), Total Hardness as CaCO3, Calcium (Ca), Magnesium (Mg), Chloride (Cl), Sulphate (SO4), Alkalinity as CaCO3 along with Sewage water analysis in terms of Biological oxygen Demand (BOD), Chemical Oxygen Demand (COD).

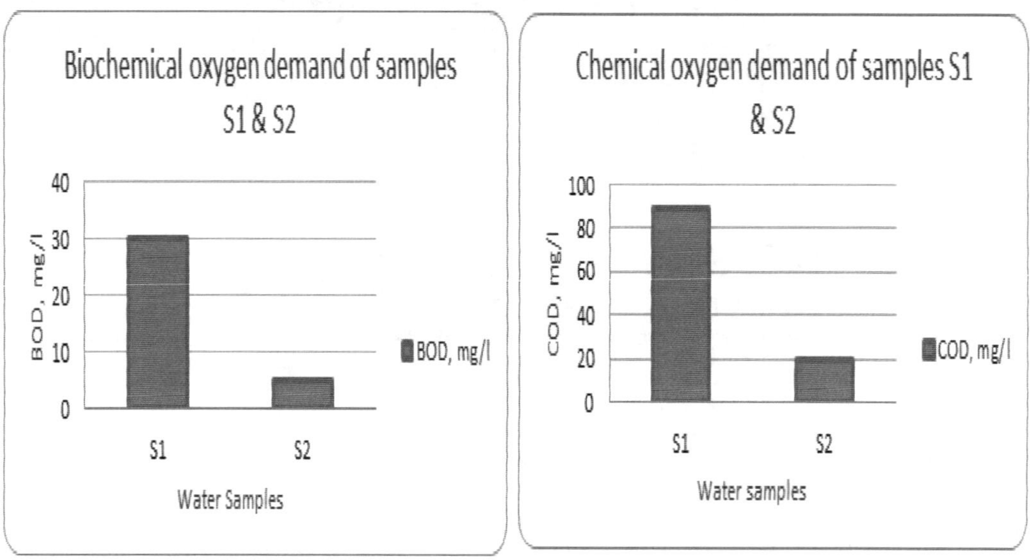

Fig 4 : BOD & COD of Sewage samples of STP

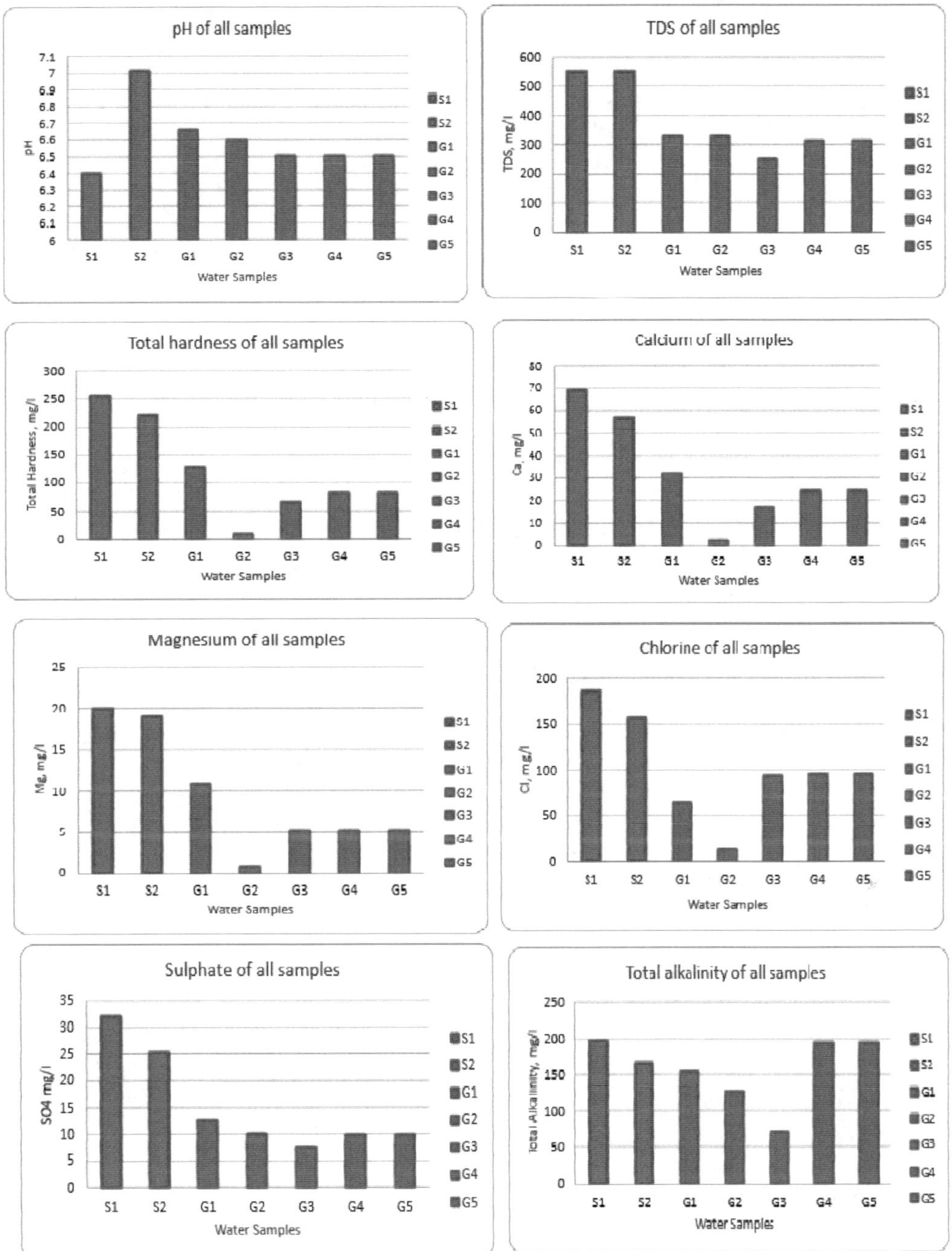

Fig. 5: Analyzed details of Sewage samples & Ground Water samples

6. Discussions & Conclusions

The analysis of the ground water and wastewater of STP (Sewage Treatment Plant) of REVA campus was done to assess its use for intended purpose. All the ground water Samples of REVA Campus are having TDS values less than 500 which is one of the required parameters for use for domestic use as per BIS 10500:1991. The other parameters tested are also within the limits of BIS. The sewage samples were analysed for BOD and COD at inlet & outlet of the STP. It was found that the BOD and COD were considerably reduced with the treatment of sewage at REVA campus before its reuse as gardening water.

The ground water of REVA Campus satisfies drinking water standards of BIS for the parameters analysed. The ground water is very much suitable for domestic use. The parameters considered for analysis of sewage from STP of REVA Campus are found to satisfy the Central Pollution Control Board (CPCB) guidelines for designated use as irrigation water and hence is very much applicable for gardening purpose & possible ground water recharge.

7. References

[1] Indu Nirupam Shukla, Kumar Rahul, Anurag Kumar, *Rainwater Harvesting – A Campus Study*,16th February 2018, www.conferenceworld.in, ISBN: 978-93- 87793-00-2, pp.290-298.
[2] S Rehan Ali, R K Jain, *Rain-water Harvesting System for College Of Engineering, Teerthanker Mahaveer University, Moradabad*, Vol. 3, Issue 8, August 2014, DOI: 10.15680/IJIRSET.2014.0308077, pp. 15649-15657.
[3] Mr. S.S. Shinde, Prof P.A. Hangargekar, *Water Conservation: Rain-water Harvesting Project For College Campus*, Volume: 07 Issue: 03 | Mar 2020, ISO 9001:2008 Certified Journal, e-ISSN: 2395-0056, p-ISSN: 2395-0072, pp. 1701-1703.
[4] Meda Kalyan Kumar, *Design of rainwater harvesting system at shilpa hostel in jntua college of engineering ananthapuramu: a case study from southern india*, Volume 11, Issue 12 (December 2015), e-ISSN: 2278-067X, p-ISSN: 2278-800X, PP.19-29..
[5] Dr sanjay D. Gaikwad, *Application of remote sensing and gis in rainwater harvesting: a case from goa, india,* volume 6, issue 1, january-2015, ISSN 2229- 5518, pp.633-639.
[6] A. S. Jasrotia, Abinash Majhi, Sunil Singh, *Water Balance Approach for Rainwater Harvesting using Remote Sensing and GIS Techniques, Jammu Himalaya, India*, 18 March 2009, DOI 10.1007/s11269-009-9422-5, pp.3035-3055.
[7] Dr Amol M Khatkhate, *Dr Sujata Rout, Case Study: Rainwater Harvesting (RWH) at Rizvi College of Engineering (RCOE)*, January 18, 2019, DOI: 10.4236/eng.2019.111002, pp.14-21.
[8] G.Elangovan, V.M. Rajanandhini, *Analysis and Design of Sewage treatment Plant: A Case Study Atnagore,* International Research Journal of Multidisiciplinary Technovation (IRJMT), 2019, 1(6); 543-552

EXPERIMENTAL INVESTIGATION ON MECHANICAL BEHAVIOR OF AAC BLOCK MASONRY

Pruthvi Raj[1], Mehar Babu Ravula[2]

1 M.Tech student, school of civil engineering, REVA University (21034014787.AKULAPRUTHVIRAJ@ce.reva.edu.in)
2 Assistant Professor, School of civil engineering, REVA University (meharbabu.ravula@reva.edu.in)

Abstract. Autoclaved Aerated Concrete (AAC) flyash Blocks are light weight bricks, which is rapidly used in high rise buildings. The AAC Blocks are one of the sustainable materials manufactured using industrial waste flyash. The masonry infills made with AAC blocks contribute to light weight of structure, thermal conductivity and faster construction. The AAC block masonry needs to be studied at material level to improve its efficiency. The material level understanding of the behavior of AAC block masonry is lacking in the literatures. In this paper, Experimental investigation is presented to understand the compressive stress-strain behavior of AAC Block masonry and its constituents. The response of AAC block masonry with weak and strong mortar is obtained from compression test.

Keywords: AAC block, Compressive strength, masonry prisms

1. INTRODUCTION

In India, the recent advances in the materials used for constructing masonry infills has got broad variation in the behaviour of the infills and its effect at the structural performance. While constructing buildings the infill walls are constructed only for the architectural purpose. The infills are considered a non-structural element and designed by neglecting its contribution in the seismic response of the structure. Infill walls typically occupy a vast quantity of material in the RC frames. Autoclaved Aerated Concrete block masonry is used as infill in the current study. Autoclaved Aerated Concrete blocks or lightweight concrete blocks are made with flyash. These are better than the burnt red clay bricks in quality and added advantages in giving comfort to the occupants. So, in most of the current structures AAC infills are used as the best alternative to clay brick infills. Heating clay brick kilns produce air pollution, and it is replaced by a steam-based heat treatment called autoclave in the AAC production industries. Also Flyash is used as one of the main raw materials in AAC manufacturing process. Hence, AAC is a sustainable building material. These Blocks were very light in weight, so it reduces the dead load to the buildings compare to other building materials. Construction with these blocks is faster and economical than normal clay bricks. AAC block masonry infills effect the overall stiffness and performance. Hence AAC masonry infill behavior has to be studied along with its prepeak and post peak behaviour. This material level understanding if once set, the response and the failure would lead to the estimate its contribution under inplane loading or out of plane loading conditions during the earthquake effect on the building.

2. LITERATURE REVIEW

In this paper, the literatures reviewed on AAC block masonry and soft clay brick masonry behavior are presented.

Bhosale et.al (2019) [1] has reported on the mechanical properties of AAC blocks and their masonry infill such as compressive strength, stress-strain behavior, and modulus of elasticity are determined experimentally. The cube specimens of sizes 50, 75, and 100 mm were cut from AAC blocks. Similarly, brick units of size 200 × 100 × 100 mm were cut to prepare the AAC block masonry. Concluded that as AAC cube increases the size, the stress decrease. AAC masonry structural properties required to model the numerical model are specified in this paper.

Amit Raj et.al (2020) [2] has performed experiments on AAC block and its masonry to evaluate mechanical properties. For the preparation of the AAC masonry prism, the unit size was taken equal to the size of clay brick. They concluded that the compressive strength of masonry is low compared to the strength of AAC block. An increase in the tensile bond strength is achieved using a high-strength mortar masonry prism than other masonry prisms.

Is 1905 -1987 [3] code of practice for structural use of unreinforced masonry code is referred to masonry structures. As per Appendix B calculate the compressive strength of masonry from the masonry prism by the aspect ratio of the masonry prism. Prisms are constructed with five brick heights and one brick width. The correction factor for various aspect ratios of masonry prism is given.

Kaushik et.al (2007) [4] The relationship between constituent properties and the compressive stress–strain relationship obtained from experiments and compressive strength of masonry made with such bricks have been investigated. Stress-strain characteristics of clay brick masonry under uniaxial compression is obtained for various proportions of cement:sand and different clay brick strengths.

Ravula and Subramaniam (2017) [5] has performed compression test on Masonry prisms made with soft clay brick with two different strengths of mortar and the behavior is presented. The failure of the masonry with soft clay brick is described based on the compression test conducted using displacement control.

3. EXPERIMENTAL PROGRAM

In this experimental program, the characterization of materials used in the masonry is presented based on the various test procedures followed at laboratory. The procedure of casting and testing is described with standard codal references. The equipment used for testing AAC Brick, mortar and Masonry are specified and experimental results obtained are presented with summary.

3.1. PREPARATION OF SPECIMEN

3.1.1. PREPARATION OF MORTAR CUBES

The mortar specimens are prepared with Ordinary Portland Cement (OPC) of Grade 53 in two strengths, strong and weak mortar M1(1:3) and M2(1:6). The Mortar specimens are prepared using the molds of 50mm cubes as per IS 4031 PART 7 [5] by clause 5.3. The compressive strength of mortar cubes is determined with an edge length of 50 mm. The W/C ratio was fixed at 0.5 for both mortar ratios M1 and M2. Ten cubes for each ratio are prepared to evaluate the strength of the mortar. By referring to clause 9.1 curing of cubes should be done by immersing in clean water for 28 days. After curing and before testing Gypsum is applied on both the top and bottom surface of mortar cubes to remove uneven surfaces for distributing the load uniformly while testing the mortar cube.

Fig.1. Curing of mortar cubes

Fig.2. Gypsum applied on both surfaces of cube

3.1.2. PREPARATION OF AAC CUBE SPECIMEN

For construction three different dimensions of Autoclaved Aerated Concrete blocks are used of dimension 100x200x600mm, 150x200x600mm and 200x200x600mm. In the present study, a 150x200x600m size Autoclaved Aerated Concrete block is used. As per IS 6441 (PART 5) 1972 [6] clause 2.1, the test specimen is prepared by cutting a cube of the edge length of 150mm with rotating blades as per the standards specifications. Three cubes are extracted from one AAC block by referring to clause 2.2 i.e one cube shall be extracted from the upper third of the sample, one from the middle, and one from the lower part of the AAC block. Before testing, A layer of Gypsum is applied on both the top and bottom surface of AAC cubes to remove uneven surfaces and to provide uniform distribution of load for testing the AAC cube.

Fig.3. Gypsum applied on both surfaces of the cube

3.1.3. PREPARATION OF MASONRY PRISMS

To construct the masonry prism the AAC blocks are cut into 200x100x70 mm same as the size of clay brick. The AAC brick masonry prisms were built with one brick along with the breadth and five bricks along with the height, using two cement sand proportions i.e strong and weak mortars. The mortar thickness of a masonry prism is10 mm. To maintain 10 mm uniform thickness, a wooden frame is made with a thickness of 10 mm, 200 mm in length, and 70 mm in width to use comfortably while constructing a masonry prism. As per IS 1905 [3], appendix B the height to thickness (h/t) or aspect ratio between 2 to 5 and a minimum height of 40 cm. In the present study, the aspect ratio is maintained 3.9 and correction factor to convert a prism strength to masonry is given in table 12 of IS 1905 [3]. The gypsum was used to cap both the top and bottom surfaces of the masonry prism. Two plywood are placed while testing between the upper and lower brick. Six masonry prisms are constructed for each mortar cube of weak and strong mortar. Curing is done for 28 days using sealed gunny bags. Masonry prism and gypsum were applied on both top and bottom surfaces of AAC bricks.

Fig.4. Masonry prism

Fig.5. Wooden frame of 10mm thickness

Fig.6. Curing of masonry prism for 28 days covered with gunny bags

3.2. COMPRESSION TESTS CONDUCTED

Compressive strength test is conducted on AAC Block, Mortar cubes of weak and strong mortar using Computerized Hydraulic controlled Universal Testing Machine (UTM). UTM can automatically capture load and displacement data to produce a graphical display of the specimen load response and it can store all the results. The sensitivity of UTM is ± 1%.

Compressive response of AAC Masonry prisms is conducted using Hydraulic Jack of 1000 kN load cell with displacement control on 4 post Loading frame. The procedures and test results are presented.

Fig.7. UTM

Fig.8. AAC Block in UTM

3.2.1. COMPRESSIVE STRENGTH TEST OF AAC BLOCK

As per IS 6441 (PART 5) 1972, to determine the compressive strength of AAC block, the specimen should be with an edge length of 150 mm. AAC block test is conducted in UTM. After testing a 6 samples, the typical graph of AAC block is presented. [6]

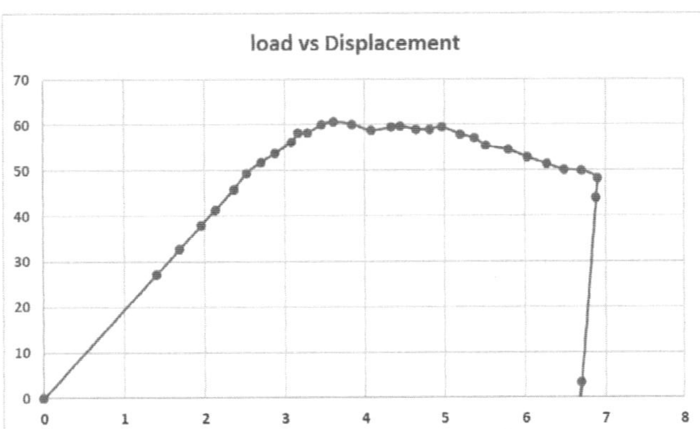

Fig.9. Load vs displacement of AAC Block

Table.1. Results of AAC cubes

Specimens	Load (kN)	Strength (MPa)
Specimen 1	60.6	2.7

Specimen 2	99.6	4.3
Specimen 3	105.8	4.7
Specimen 4	95.0	4.5

3.2.2. COMPRESSIVE STRENGTH TEST OF MORTAR CUBES

As per IS 4031 PART 7 [5] to determine the compressive strength of mortar, 50mm cube is tested in the present study. Strong and weak mortar cube strength is obtained. The load response of weak and strong mortar cubes are presented in the Fig. 10 and the results are consolidated in Table 2.

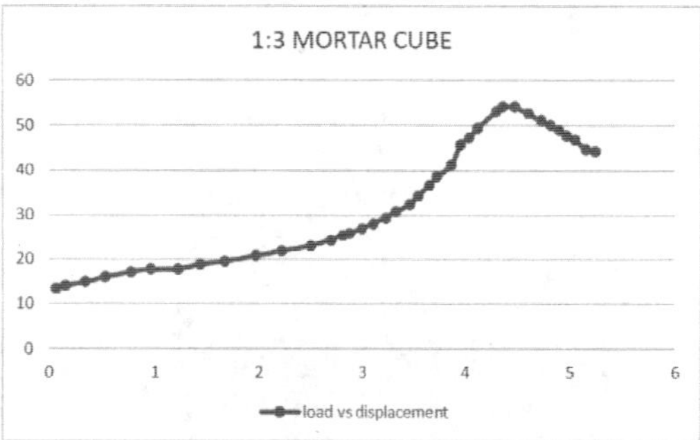

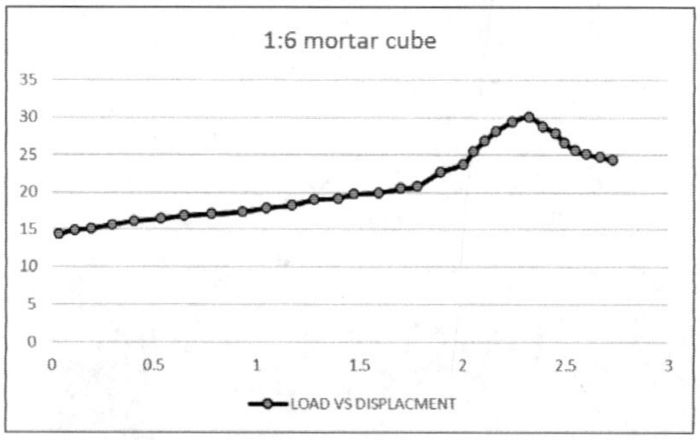

Fig.10 Typical load response of 50mm Mortar cube c:sa (1:3) and (1:6) by volume

Table.2. Results of 1:3 and 1:6 Mortar cubes

	1:3 Mortar cubes		1:6 Mortar cubes	
Specimens	Compressive Load (kN)	Compressive strength (MPa)	Compressive Load (kN)	Compressive strength (MPa)
Specimen 1	53.83	21.53	30.125	12.05
Specimen 2	54.12	21.64	21.43	8.57
Specimen 3	49	19.6	28.28	11.37
Average	52.31	20.92	26.61	10.66

3.2.3. COMPRESSION TEST ON MASONRY PRISM

Masonry prisms made with AAC Bricks, mortar of 1:6 (c:sa), 1:3 (c:sa) are casted and tested under compression with displacement control using a hydraulic jack and load cell, of capacity 1000kN mounted on a four post Loading frame. The measurement instruments used are two LVDT's of 100mm maximum displacement, and 5mm Strain Gauges. The test setup in the loading frame is shown in the Fig. 11

Fig.11. Setup in the loading frame

3.3. RESULTS AND DISCUSSION

3.3.1. COMPRESSION TEST ON AAC MASONRY PRISM

After curing for 28 days, AAC masonry prism is tested in loading frame under constant application of displacement. For strong mortar (1:3-c:sa), a crack initiation is observed at a load of 280kN and for weak mortar masonry prism, a crack is observed at 220 kN for 1:6 (c:sa) Masonry Prism. Early cracking in the weak mortar masonry prism. Typical Stress-Strain response of the masonry prism of strong and weak mortar strengths is presented in the Fig. 12. Summary of the strengths obtained from the compression test on strong and weak mortar masonry prisms is presented in Table 3 and the specimen showing incomparable strength is not considered while averaging so as to avoid errors while testing procedure. The correction factor is applied for the masonry prism and obtained masonry strengths are presented in Table 4. There is not much variation in the strengths of both the prisms but the failure in 1:6 has shown a difference.

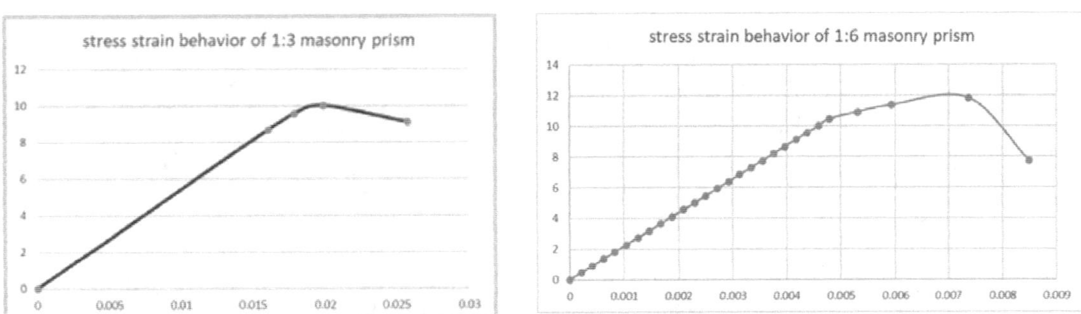

(a) (b)

Fig.12. Typical Behavior AAC masonry prisms

Table.3. Prism Strength and E modulus of AAC masonry

	Masonry 1:3 (c:sa)		Masonry 1:6 (c:sa)	
Specimens	PRISM STRENGTH (MPa)	E MODULUS (MPa)	PRISM STRENGTH (MPa)	E MODULUS (MPa)

Specimen 1	12.27	977	11.8	2176
Specimen 2	12.752	919	11.8	930
Specimen 3	10	1002.5	10.45	636
Average	11.66	966.166	11.125	783

Table.4. Masonry Strengths table obtained by applying correction factor to prism

	Prism strength (MPa)	Correction values as per IS 1905	Masonry strength (MPa)	E modulus (MPa)
1:6	11.125	0.942	10.47	783
1:3	11.66	0.942	10.98	966.1

3.3.2. MASONRY FAILURE:

In both the masonry prisms, the crack initiated in the mortar joint and penetrated into the brick leading to failure of the specimen. 1:6 (c:sa) masonry prism failed with spalling and multiple cracking in the mortar joint. The crack penetrated through the brick forming a single major crack leading to failure of the masonry prism 1:6 (c:sa) as shown in Fig. 13. In 1:3 (c:sa) masonry prism, cracking load is near to the ultimate load, and one single crack is formed leading to failure of the masonry prism. The crack propagation is very fast in 1:3 (c:sa) where the difference between the point of nonlinearity and the ultimate load is very less and the specimen with 1:3 (c:sa) showed brittle failure,

(a) (b)

Fig.13. (a) Spalling and multiple cracking in 1:6 (c:sa) masonry specimen and **(b)** single cracking in 1:3 (c:sa) masonry specimens leading to Failure.

4. CONCLUSION:

- The strength of AAC brick obtained from compression test results is 4.5MPa which is in the lower range of regular clay bricks listed in IS1905.
- The Elastic modulus obtained is 345 MPa which shows a very soft material compared to other materials of brick units.
- The Mortar strengths obtained for 1:3 (c:sa), and 1:6 (c:sa) are 20.9 MPa and 10.6 MPa respectively which is higher than strength of AAC brick.
- The compressive strength obtained for 1:3 masonry prism is 11.66 MPa and Elastic modulus 966.6 MPa and Masonry prism strength obtained for 1:6 masonry prism is 11.35 MPa and E modulus 783 MPa. The strengths are nearly comparable inspite of different mortar strengths and Elastic Modulus.
- The Stress-Strain response is obtained for 1:3 and 1:6 masonry prism made with AAC Blocks, showed nearly early nonlinearity in 1:6 compared to 1:3.
- The failure pattern for 1:3 and 1:6 masonry prism is identified as multiple cracking in 1:6 along with spalling of AAC brick whereas one single crack is observed for 1:3 masonry prism.
- Failure of masonry prism 1:6 is initiated in the mortar joint and penetrated into the AAC brick.
- In 1:3 (c:sa) masonry prism, cracking load is near to the ultimate load indicating the crack propagation is very quick in 1:3 (c:sa) indicating brittle failure.
- In 1:6 (c:sa) masonry prism, the cracking load and the ultimate load has showed some lagging indicating multiple cracking and spalling until load reaches the ultimate.

5. REFERENCES:

[1] A. Z. N. P. D. R. &. S. P. Bhosale, "Experimental Investigation of Autoclaved Aerated Concrete Masonry," American Society of Civil Engineers, vol. 21(3), no. 10.1061/(ASCE)MT.1943-5533.0002762., pp. 1-11, 2019.

[2] A. C. B. &. U. S. D. Amit Raj, "Evaluation of Mechanical Properties of Autoclaved Aerated Concrete (AAC) Block and its Masonry," Journal of The Institution of Engineers (India): Series A, vol. 101, no. 12 march 2020, pp. 315-325, 2020.

[3] Is 1905 -1987 code of practice for structural use of unreinforced masonry, INDIA, 1987.

[4] Kaushik HB, Rai Durgesh C, Jain Sudhir K (2007) Stressstrain characteristics of clay brick masonry under uniaxial compression. J Mater Civil Eng 19(9):728–739

[5] IS 4031 PART 7, 1996, Method of physical test on Hydraulic cement.

[6] IS 6441 (PART 5) 1972, Methods of test for autoclaved cellular concrete pro.

[7] Ravula, M. B., and K. V. L. Subramaniam. 2017. "Experimental investigation of compressive failure in masonry brick assemblages made with soft brick." Mater. Struct. 50 (1): 19

IOP Conf. Series: Materials Science and Engineering 1255 (2022) 012017 doi:10.1088/1757-899X/1255/1/012017

An Experimental Investigation on Characteristic Properties of Hybrid Fibrous Ferro Cement

Chandrashekar A R[1*], **Dr. Kannam Praveen**[2]
*1,Research Scholar, 2 Associate Professor, School of Civil Engineering, REVA University, Bengaluru – 560064, Karnataka, India. (*R21PCV04@reva.edu.in, kannam.praveen@reva.edu.in*)*

Abstract. Ferro-Cement is a notably new material such as cord meshes and cement mortar. This fabric turned into developed via P L Nervi, an Italian architect and engineer, in1940. He revived the idea of Ferro-Cement and used it for building fishing boats. Ferro-Cement is skinny and light weight cement composite, with numerous specific houses. The inclusion of fibers to the concrete can amplify the energy homes of the blended and transfigure the frangible cement blended into more consistent and malleable fabric known as FRC. Yet the both substances consists few draw backs. Both the substances unable to be waged where unreasonable tremble, unreasonable bending intensity and excessive effect are to be endured. Accordingly the answer lies within the new compound known as fibrous Ferro-Cement, that's the mixture of fibber bolstered concrete and fibrocement. This new composite has some stepped forward mechanical homes, inclusive of sturdiness, impact resistance and so on. Addition of a couple of kind of fibers to the composite improves the energy houses to even greater quantity. However the inclusion of multiple kind of fiber in to the fibrous Ferro-Cement steers to hybrid fibrous Ferro-Cement, which is having more power, homes than the mono fibrous Ferro-Cement.

1. Introduction

Concrete is a composite cloth composed of excellent, coarse materials bonded together with a cement fluid and hardens over a time. Concrete is second most broadly used fabric after water, and most consuming constructing materials. Concrete structures that have been built around the arena are situation to a huge variety of various conditions of use and acquaintance to environmental conditions comprising erosion, climate, and pollution. All environmental situations observed concrete as the iconic and the fine binding material. Besides regardless of the situations might be, concrete is anticipated to provide nice overall performance for the wholeness of their carrier life with little or no care.

Concrete, although sturdy in compression, is susceptible in tension. For this reason it wishes assist in resisting tensile stresses as a result of bending forces from implemented masses which could result in cracking and in the long run failure. In this situation the metallic reinforcements are embedded in concrete member. But on the same time it will increase with self weight. In order to make it lessen fibrous ferrocement is the first-class alternative.

Fibrous ferrocement which is a aggregate of fibre bolstered concrete and ferrocement combines the blessings of fibre bolstered concrete and ferrocement and is a premising cloth. . Ferrocement and fibre strengthened concrete (FRC) had a long records in civil engineering discipline, now an afternoon's aggregate of these two i.e., fibrocement is popular. In this test we're studying flexural and compressive behavior of monofibres and polyfibres in ferrocement with distinct extent fraction. The addition of fibres in the matrix has many essential outcomes on Fracture power, Toughness, effect resistance,

Flexural strength resistance. Fibres used in the ferrocement act as secondary reinforcement and arrest the cracks forming and propagating. It will increase compressive strength and flexural power of ferrocement, enhances the elastic modulus and decrease brittleness.

2. Standard Materials Used

2.1 Cement

OPC of grade 53 confirms to IS 12269 – 2013 is used for this investigation. Specific Gravity of cement is 3.15 and a standard consistency of 30%.

2.2 Fine Aggregates

M Sand passing through the IS Sieve 4.75mm and retained in 75μ confirmed and classified to Zone II according to IS 383-1970 selected and used for specimen preparation.

2.3 Poly Propylene Fibers

Polypropylene fibres from textile industries having aspect ratio 400 and density 930kg/m^3 are used.

2.4 Steel Fibers

Steel fibres of end hooked type having aspect ratio 50 and density 7850 kg/m^3 are used in this investigation.

2.5 Hexagonal Wire mesh

It is also normally referred to as bird twine mesh is product of bloodless drawn wires of diameter 16 gauge and woven into hexagonal styles with mesh beginning varying from 10 mm to twenty-five mm This is the most inexpensive, simplest to address and maximum typically used in Ferro-cement creation.

2.6 GI Mesh

Welded mesh of beginning 12.5mm×12.5mm (1/2inch) is incorporated in the periphery. The welded mesh have small diameter bars (16 gauges) kept closely in each guidelines and are spot welded .It is fabricated in rectangular or rectangular sample with the aid of perpendicular intersecting wires (generally 2-3 mm diameter) product of low to medium tensile power metal (which can be a great deal stiffer than hexagonal or woven wire mesh) and welded together on the intersections.

3. Methodology

A cement and sand mortar with a share of one: three become organized with a water cement Ratio of 0.4. Casting of specimen following process is adopted for casting the specimens,

- Calculated quantity of cement and sand is weighed and dry blended very well to get uniform blend.
- Calculated amount of water is added to the dry combo to get uniform mix, moulds are cleaned and gently oil is done to the inner floor of the mould.
- The cages prepared thru mesh are first placed in the mould after which fibres are mixed with mortar at the time of blending and it's far poured within the mould.
- An Experimental Investigation on the Characteristic Properties of Hybrid Fibrous Ferro

cement.
• All moulds have been kept on desk vibrator and sufficient vibration becomes given to compact the fibrous mortar.
• The specimens had been completed clean after vibration. After 24 hours of casting, the specimens are demoulded and saved in water tank for Curing for 28 days.

4. Results and Discussions

After curing specimens in potable water and their mechanical strength parameters were tested.

4.1 Compressive strength

Cubes of size 150*150*150 mm were casted and cured for 7 & 28 days and tested for compressive strength. Following results are obtained after testing the cubes.The Compressive strength of cement mortar with fibre and mesh skeleton shows better performance compared to conventional. The increment in the strength is due to the proper orientation of steel and polypropylene fibres. Strength comparison with cement mortar with mesh gives 31% more compressive strength than plain cement mortar.

Table 1 Compressive Strength (C.S) Results of 7 and 28 days

Mix	C.S for 7 days(Mpa)	C.S for 28 days(Mpa)
Only Mortar	17.03	18.22
With Mesh	16.15	18.51
PPF	21.18	22.25
SF	22.96	24.14
SF+PPF	25.03	27.18

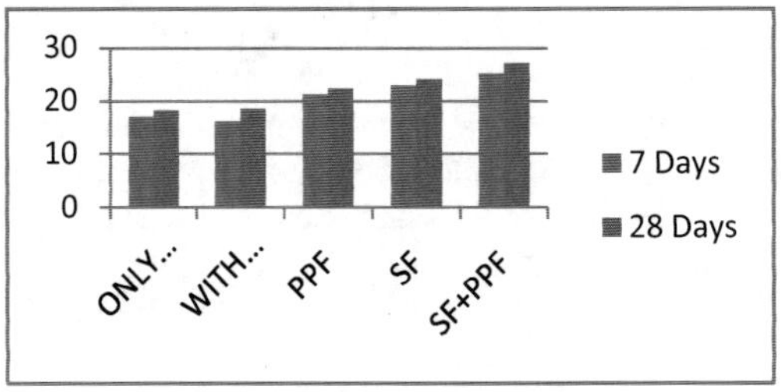

Fig 1 C.S results of 7 and 28 days

4.2 Flexural Strength

Flexural Strength of cement mortar with fibre and wire mesh skeleton shows better performance. The increment in the strength is due to the proper orientation of steel and polypropylene fibres. Strength comparison with cement mortar with mesh gives 61% more compressive strength than plain cement mortar.

Table 2 Flexural Strength (F.S) Results of 7 and 28 days

Fibre description	7 days (MPa)	28 days (MPa)
Only Mortar	2.25	4.25
With Mesh	5.75	6.25
PPF	9.25	9.41
SF	11.83	11.95
SF+PPF	11.29	12.29

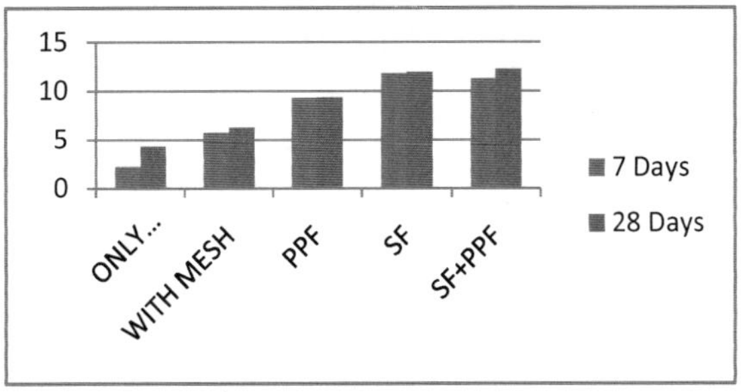

Fig: 2 Flexural Strength (F.S) Results of 7 and 28 days

5. Conclusions

In the present investigation the fibres are

- Compared to traditional cement mortar the compressive energy and Flexural electricity of cement mortar with mesh has been expanded by using 1.Fifty nine% and forty seven.05% for 28 days
- Compared to cement mortar with mesh the compressive power and Flexural power of cement mortar with PPF has been multiplied with the aid of 20.2% and 50.Fifty six% for 28 days
- Compared to cement mortar with mesh the compressive strength and Flexural energy of cement mortar with SF has been multiplied via 30.4% and 91.2% for 28 days
- Compared to cement mortar with mesh the compressive energy and Flexural energy of cement mortar with Hybrid fibres(SF+PPF) has been accelerated by way of 46.8% and ninety six.Sixty four% for 28 days
- Compared to cement mortar with PPF the compressive electricity and Flexural strength of cement mortar with Hybrid fibres(SF+PPF) has been improved with the aid of 22.15% and 30.60 % for 28 days
- Compared to cement mortar with SF the compressive power and Flexural energy of cement mortar with Hybrid fibres (SF+PPF) has been elevated through 12.Fifty nine% and 2.Eight% for 28 days.

6. References

- ChandraShekhar Rao T, Gunneswara Rao T D, Ramana Rao N V and Rambabu C H,"An experimental study on ferrocement box-beams under flexural loading" International Journal of Emerging Technology and Advanced Engineering(IJETAE) ISSN:2250-2459, Vol 2, Issue 9, September 2012.

- Deepa S, Thenmozhi R, "An experimental investigation on the flexural behaviour of SCC ferrocement slabs incorporating fibres" International Journal of Engineering Science and Technology (IJEST), Vol 4, Issue 5, May 2012.

- Murali G, Arun E, Arun Prasadh A, Infant raj R and Ashwin Prasanth T, "Experimental investigation of reinforced ferrocement concrete plates under impact load" International Journal of Latest Research in Engineering and Computing(IJLREC) ISSN:2347-6540,Vol 2,Issue 1,PP 01-04.

- Mohammed Arif, Pankaj and Surendra K.Kaushik, "Mechanical behaviour of ferrocement composites: An experimental investigation" Cement and Concrete Composites(ELSEVIER),Vol 21(1999), PP 301-312.

- Patil N K and Prakash K B "Study on effect of replacement of cement by microsilica on compressive strength of fibrous ferrocement produced with GI fibers", Structural Engineering Digest, Jan-March 2010.

- Prahallada M.C and Prakash K.B, "Characteristics of hybrid fibre reinforced concrete (HFRC)-An experimental study " Journal of Civil, Structural, Environmental, Water resources and Infrastructure Engineering Research (JCSEWIER) Vol 2, Issue 2, June 2012, PP9-15.

- Ravi K and Prakash K B, "Strength characteristics of hybrid fibre reinforced self compacting concrete subjected to sustained elevated temperature", 33rd conference on our world in concrete and structures: 25-27 August 2008,Singapore.

- Sakthivel.P.B and Jagannathan.A, "Ferrocement construction technology and its applications-A review" International Conference on Structural Engineering, Construction and Management(ICSECM): 15-17 December 2011,Srilanka.

- Shannag M J and Mourad S M, "Flowable high strength cementitious matrices for ferrocement applications" Construction and Building Materials (ELSEVIER),Vol 36(2012), PP 933-939

- Sudhikumar G S, Prakash K B and Seshagiri Rao M V, "Effect of freezing and thawing on the strength characteristics of slurry infiltrated fibrous ferrocement using steel fibres", Global Journals of Research in Engineering: E (GJRE-E), Vol 14, Issue 5, 2014

Strengthening of Cement mortar by Reinforcing Carbon Based Nano Material

Jayachandra[1], H. J. Yashwanth[2], Dr. Y Ramalinga Reddy[3]*, Sanjith J[4]
[1]*Research Scholar & Assistant Professor, School of Civil Engineering, REVA University;, Bengaluru – 560064, Karnataka, India.*
[2]*Assistant Professor, Acharya Institute of Technology, Soledevanahally, Bengaluru-560107.*
[3]*Professor & Director, School of Civil Engineering, REVA University, Bengaluru – 560064, Karnataka, India.*
[4]*Assistant Professor, Department of Civil Engineering, Adichunchanagiri Institute of Technology, Chikkamagaluru, Karnataka-577101*
(jayachandrajc65@gmail.com; yashwanthhj95534@gmail.com ; dir.ce@reva.edu.in; sanjugou@gmail.com)

Abstract. The present study reported the mechanical behaviour of reduced Graphene Oxide infused in cement mortar. In this investigation initially well graded Graphene Oxide confirming the presence of carbon functional group by FTIR Spectro results it is dispersed in suitable dispersion agent to obtain reduced Graphene Oxide [rGO]. The reduced Graphene Oxide was ultrasonicated with different hours of sonication at 1, 2, 4 and 6hours to reduce the oxygen functional groups. A series of cement mortar composite with varying percentages of reduced Graphene Oxide were prepared to achieve maximum strength by examining the compressive strength for different curing periods. The mechanical test results revealed that the composite containing reduced Graphene Oxide with 0.1% dosage showed good dispersion and well compressive strength prepared for 28 days than the plain cement mortar composite with 4hour sonication time. The strength of cement mortar prepared decreased after the time of sonication amplified from 4hours to 6hours due to interfacial bonding between the cement paste decreased. This study delivers a valued influence toward better understanding of the influence of reduced Graphene Oxide particle size on the properties of cementitious composites to optimize their performance.

Keywords: reduced Graphene Oxide [rGO], X-ray diffractogram, Raman Spectra, Ultrasonication, Compressive strength.

1. Introduction

Cement is the most important and one among widely used construction materials in the society due to its economical, good physical, chemical and mechanical properties. Nonetheless, because of its low tensile strength, heat of hydration, and strain capacity, it is only suitable for structural applications that do not require reinforcement. Tensile strength is low owing to pre-existing defects. Using reinforcement such as steel, fibres, and so on, it is possible to prevent the emergence of micro cracks and therefore enhance tensile stress resistance. Recent advancement in nanoscience and nanotechnology was produced many nanomaterials like TiO_2, ZnO. Go etc for many application like sensors, electronic devices, hydrogen generation. Reinforcement in concrete-based material. Introducing nanomaterial in cement based nanocomposites helps improve the hydration process. This motivated to study the reinforcement of nanomaterials in cement-based material, also by modifying their structure at nanoscale it will helps in improving their macro structure and mechanical properties of the cement. Recent studies revealed that extensive research is being conducted on the reinforcement of carbon-based materials such as graphene Oxide (GO), reduced Graphene Oxide (rGO), Carbon Nano Tubes (CNTs), and Multi Walled Carbon Nano Tubes (MWCNTs) in cement-based nanocomposite and their properties are being reported. These might be utilized as reinforcements in inhibit the formation and delay the propagation of micro cracks . Nano fibres have been found to be more effective than ordinary steel fibres in improving the mechanical performance of cementitious materials because they help to restrict cracks before they develop into micro cracks. The use of Nanoparticles contributes to making of small crystals such as Ca $(OH)_2$ and constant agglomeration of C-S-H products. This will also contribute in accelerating pozzolanic reactions which consumes Ca $(OH)_2$ and produce added C-S-H gel which helps in improving the contact zone which improves the bonding strength between cement compound and aggregate . Graphene has been known for over a century, with Kphi Scutters publishing his first paper in 1917 describing Graphene Oxide. It was not widely recognized, however, until Novoselov and Geims demonstrated mechanical exfoliation of graphene films in 2004. The surface of GO contains functional groups such as epoxide, carboxyl, hydroxyl, and carbonyl, and it disperses well in the cement matrix. Many researchers performed molecular dynamic analysis to calculate the interfacial bond strength between calcium silicate hydrated (C-S-H) gel and carboxyl-functionalized graphene nanoplates and discovered that the interfacial bond strength between pristine, hydroxyl-functionalized, amine-functionalized, and carboxyl-functionalized Graphene nanoplates and this helped in the positive impact. This indicated in Graphene nanoplates could enhance the interfacial strength, mechanical properties, and toughness of cement composites.

2. Materials and Properties

2.1 Materials:

1. Cement(OPC53 Grade)
2. Fine aggregate(M-Sand)
3. reduced graphene Oxide (rGO)

Binder: Ordinary Portland Cement (OPC) confirming to Indian standard code IS 12269-1987.

Table1. Properties of cement

Physical properties	Test results	IS Limits (IS: 8112-1989)
Specific gravity	3.11	3.15
Initial setting time	52 minutes	Not less than 30 minutes

Fine aggregate: Manufactured sand (M-sand) conforming to Indian standard code.

Table2. Properties of fine aggregate

Properties	Experimental Results
Specific gravity	2.34
Grading Zone	Zone II

2.2 Physical properties of rGO

Figure1. (a) & (b) SEM image of as synthesized rGO

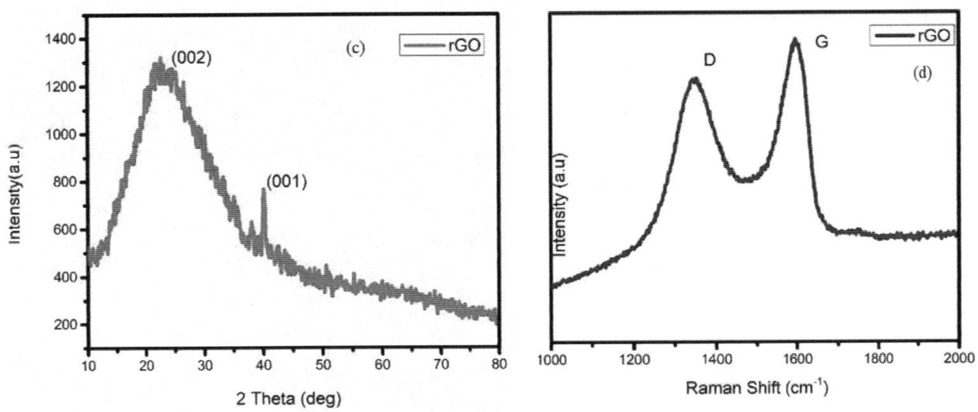

Figure1(c) X-ray diffraction pattern (d) Raman spectra of as synthesized rGO

The size, morphology, and crystal phase of the as synthesized rGO was studied using Scanning electron microscope. Figure 1(a) & (b) shows SEM images of as synthesized rGO, it is understood rGO is sheets like structure. Further in order to study crystal structure as synthesized rGO was characterized using X-ray diffractogram results are depicted in Figure 1(c). Which shows broad peak at an angle 22^0 corresponds to the amorphous graphitic carbon plane (101) which is good agreement with previously reported literature. Further Figure 1(c) shows the Raman spectra of as synthesized rGO sheets which two peaks around 1362 cm^{-1} and 1586 cm^{-1} respectively corresponding to D band and G band respectively. Where D band arises due to the k-point breathing mode of A_{1g} phonons, and G band which arises due to the tangential stretching mode of E_{2g} phonons of sp^2 atoms. This SP^2 hybridisation can improve the hydration of cement.

2.3 Methodology

2.3.1 Preparation of rGO

Chemicals are purchased from sigma aldrich with laboratory grade. Modified hummers method was adapted to synthesis Graphene Oxide (GO). In brief, 0.5g of sodium nitrate and graphite powder was dissolved in the 23ml of sulphuric acid kept it in ice bath and stirred it for 30mins then 3.0g of potassium permanganate slowly added to the solution at 20^0c. Then the obtained solution was transferred into water bath and stirred it for an hour until the paste is obtained. Further 100ml of water added to the solution and heated at 90^0C under continuous stirring for 15minutes. Then the obtained solution is diluted with 500ml of water and 3ml of H_2O_2 until the colour of the solution change from dark brown colour to yellow colour. The obtained mixture further purified using water and ethanol using filtration process. The obtained product was dried in vacuum and used for further preparation of reduced graphene oxide. As synthesized GO was dispersed in water with the concentration of 1 mg mL^{-1} and sonicated for one hour to get rGO.

Figure2. As prepared reduced Graphene Oxide (rGO)

2.3.2 Preparation of rGO/cement mortar

Normal Portland cement and Fine aggregates (Sand) confirming to Indian Standards was used in the composition of mortar mix. Moulds of size 70.6 * 70.6 * 70.6 mm cube were used. Cement: fine aggregate ratio has been taken ass 1:3 with 0.5 water cement ratio, Then different weight percentage of rGO (0.02%, 0.04%, 0.06%, 0.08%, 0.1%) was added separately and prepared the mortars and studied their properties.

Figure3. Prepared mortar cubes

2.3.3 Mechanical Compressive Strength.

Compressive tests were performed on the as prepared different weight of rGO reinforced samples at 7, 14 and 28 days after curing to investigate the mechanical enactment of the composites.

3. Results and discussions

3.1 Cube Compressive strength of rGO/Cement mortar

Compressive strength of the as prepared mortars with different weight percentage of rGO (0.02wt% to 0.1 wt%) was studied for 7, 14-days and 28-days and the outcomes are depicted in figure 4. Which illustrates as increased amount of reduced Graphene oxide in the cement mortar the compressive strength was increased gradually then maximum compressive strength obtained 34 n/m for 0.1wt% of rGO as compared to plane cement mortar. This is due to increase in the charge density, more hydrophilicity nature provided the rGO sheets which helps to improve in hydration of cement hence it will increase the strength of the cement

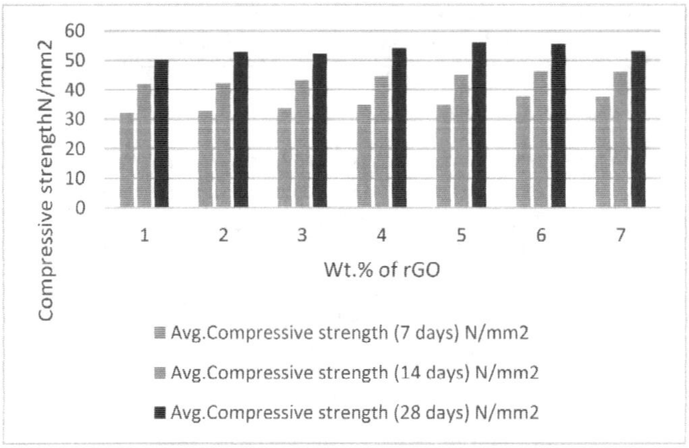

Figure4. Compressive strength test result of rGO Cement composite.

Table 3: Compressive strength result of rGO reinforced cement mortars results

Wt.% of rGO	7Day's Compressive strength N/mm^2	14Day's Compressive strength in N/mm^2	28Day's Compressive strength N/mm^2
0	32.09	42.10	50.02
0.02	32.12	42.23	52.23
0.04	32.35	42.53	52.24
0.06	31.60	43.78	53.15
0.08	33.45	45.86	54.76
0.10	34.21	46.50	55.63
0.12	34.03	46.11	53.23

From above test results the compressive strength of cement mortar composites with different weight percentage of rGO was studied. It is found that there is an improvement in compressive strength with the incorporation of rGO with different dosage in weight percentage. The optimum result was obtained at 0.1wt% of rGO incorporation in cement mortar further increasing the amount of rGO in the cement decrease in the compressive strength

Conclusion

To improve cement's mechanical properties rGO is infused in cement mortar of varying percentages and to ensure it, compressive strength was conducted. The compressive strength performance is shown in Figure 4 for varying percentages of rGO and for different curing periods, rGO of 0.1% infused in mortar specimen showing maximum compressive strength for all 7, 14 and 28 Days curing period which proves that rGO is capable in enhancing the mechanical properties of cement composites because of hydration process takes place during curing process.

Conclusion

Based on the above study reduced graphene oxide can be used as cement composite in mortar mix.

- Reduced graphene oxide with 1, 2, 4 and 6 hours of sonication reduced the oxygen functional groups which abetted in refining the hydration of cement composite.
- The general pattern showed that there was an upsurge in the compressive strength from 7, 14 and 28 days and the maximum compressive strength obtained for cement mortar with 4-hour sonication.
- By bring together rGO into cement composite there is an improvement in molecular bonding, hydration degree and crystallinity of the composite.
- However further intensification in the rGO particle size had negatively affected on the molecular bonding above 4hour of sonication.

Acknowledgements

I would like to acknowledge my Research supervisor for his support and REVA University for providing all the facility which has always been a tremendous source of inspiration.

References

[1]. Colin R. Gagg.,(2014). Cement and concrete as an engineering material: An historic appraisal and case study analysis. http://dx.doi.org/10.1016/j.engfailanal.2014.02.004.

[2]. Song Wang, Shupin Zhang, Yaping Wang, Xiaoning Sun and Kangning Sun, Reduced Graphene Oxide/Carbon Nanotubes Reinforced Calcium Phosphate Cement, Ceramics International, http://dx.doi.org/10.1016/j.ceramint.2017.06.196

[3]. Chuah, S., Pan, Z., Sanjayan, J. G., Wang, C. M., & Duan, W. H. (2014). Nano reinforced cement and concrete composites and new perspective from graphene oxide. Construction and Building Materials, 73, 113-124.
[4]. Al-Rifaie, W. N., & Ahmed, W. K. (2016). Effect of nanomaterials in cement mortar characteristics. J. Eng. Sci. Technol, 11(9), 1321-1332.
[5]. Cao, Y., Zhang, J., Feng, J., & Wu, P. (2011). Compatibilization of immiscible polymer blends using graphene oxide sheets. Acs Nano, 5(7), 5920-5927.
[6]. Enhanced acidity and pH-dependent surface charge characterization of successively oxidized graphite oxides.
[7]. Suk, J. W., Piner, R. D., An, J., & Ruoff, R. S. (2010). Mechanical properties of monolayer graphene oxide. ACS nano, 4(11), 6557-6564.

Performance of Concrete made with Natural Aggregate and Multi-cycle Recycled Concrete Aggregate

Madhu K A [1*], Manoj V[2], Dr.Rajakumara H N [3],
1,2, Assistant Professor, School of Civil Engineering, REVA University, Bengaluru – 560064,
3, Professor and Head, Department of Civil Engineering BMSIT Karnataka, India.
(madhu.ka@reva.edu.in ; manojvchandra90@gmail.com; drhnrajkumara@bmsit.in)

Abstract. The research concentrated on the use of concrete demolition waste (CDW) to achieve long sustainability, as coarse aggregate in the production of Recycled aggregate concrete (RAC). Concrete demolition waste (RCA) is used to replace natural coarse aggregates (NCA) in M-25 grade concrete production, preserving natural resources and reducing the amount of demolition waste that must be landfilled. The effect of substituting 25%, 50%, and 100% NCA with three cycles of RCA on the mechanical characteristics with an endurance of hardened concrete was studied and compared to control concrete. The results of this study show that it is possible to use the RCA coming from CDW in order to the production of concrete. The performance of concretes with 0% and 25% replacement of NCA by RCA was nearly identical to that of concrete without replacement, but the strength of concrete was reduced with 50% and 75% replacement.

Keywords: concrete demolition waste, concrete demolition waste, Recycled aggregate

1. Introduction

Sustainable construction is presently turning into a significant concern in terms of construction development, to the detriment of our planet's future. The aggregate is one of the most critical parts of concrete's strength. However, the production of concrete with NCA has its own environmental impact due to the depletion of natural resources [1]. Every year, more than 3 billion tonnes of CDW are generated in 40 major nations throughout the world [2]. The overconsumption of natural resources is the major cause of environmental degradation, destruction, and decline of natural and mineral resources. Aggregates make up around 60% to 75% of total concrete volume [3]. As a result, it is critical to reduce natural resource use and recover solid waste. Researchers have proposed several techniques to pre-treat RCA prior to use in concrete, including immersion in acid, carbonation, nano-silica treatment, and pozzolana slurry, because of the detrimental impact of RCA on concrete is related to the old-adhered mortar [4]. RCA has higher water absorption values. As a result, RCAs are considerably more permeable than natural ones [5]. Increased demand for natural resources for concrete manufacture has a considerable impact on mountains and rivers as a supply of non-renewable minerals [6]. Over 90% of CDW is presently disposed of in landfills; however, it can potentially be recycled [7]. Recycling aggregate in concrete is a possible solution to the problem of CDW [8]. Recycling concrete not only reduces the demand for NCA, but it also reduces the amount of waste that ends up in landfills [9]. Using RCA obtained from crushing demolition or earthquake waste concrete for fresh concrete production is one of the key approaches to constructing a sustainable construction [10]. The fundamental distinction between RCA and NCA is the mortar that is applied to RCA [11]. A global R&D effort has proven the

potential use of RCA as a concreting material during the last several decades. Researchers have recommended that more water be put into concrete to fulfil the water absorption requirements of RCA [12]. The construction sector produces 40% of the world's trash. India's central pollution control board estimates that solid waste production is about 48 million tonnes per year, with the construction industry responsible for 25% of it [13].
Increased demolition volume from ancient buildings, concrete pavements, and other structures that have outlived their usefulness owing to structural deterioration beyond repair and must be removed [14]. The quality of RCA is determined by the quality and source of the CDW from which the RCA is created 15]. Due to the high water-cement ratio, the utilization of extra mixing water in RAC lowers the macro characteristics of RAC [16]. The use of RCA to produce RAC has the potential to minimize environmental impact while increasing construction sustainability [17]. The use of RCA in the manufacture of fresh concrete can help to manage the rising solid waste problem while also preventing the depletion of natural aggregate supplies [18]. Semi-saturation of the RCA prior to inclusion can result in self-compacting concrete with the required workability and strength [19]. Larger weight losses were recorded between 1000 – 3000 C in contrast to conventional concrete, owing to increased water loss in more porous recycled particles [20]. The inclusion of RA reduces the durability of concrete. The inclusion of RAs increases the shrinking of SCCs. Instead of natural aggregates, a partial or complete replacement may be a viable low-cost alternative for the composition of concrete that fulfils quality and durability criteria. This is now a hot issue among building professionals. The stress-strain ratio fluctuation of RAC was comparable to that of NAC. Sustainable concrete is based on the replacement of stone aggregates of natural origin for construction and demolition waste, and it proves to be Sustainable concrete is based on the substitution of natural stone aggregates for CDW. According to a review of the current literature, limited information is available on concrete containing RCA. As a result, a comprehensive investigation has been conducted to see if RCA can be used as a partial replacement for NCA in the production of M25-grade concrete. Three consecutive RCA (25, 50 and 100 %) by volume are obtained and have been used as a partial substitute for NCA.

2. Materials and Methods

Cement: As a binder, ACC cement adhering to IS: 8112: 1989 was used. It has a surface area of 3220 cm2/g and a specific gravity of 3.14.
Fine Aggregate: M-sand that passed through a 4.75 mm IS sieve according to IS: 383-1970 grading Zone-II was utilized. The incorporated sand has a fineness modulus of 2.20 and a specific gravity of 2.62.
Coarse Aggregate: There are four different types of coarse aggregate used in this study, namely NCA, RCA1, RCA2, and RCA3. Natural course aggregates (NCA) are collected from a local source and used in the control mix and recycled concrete aggregate (RCA) was collected from the crushing of the concrete specimens. RCA was collected in the IS sieve size of 20 mm passing but 4.75 mm retaining according to IS: 383-1970 grading Zone-II was utilized. Table 1 lists the physical characteristics of aggregates. Figure 1 shows the crushing of concrete samples to collect the recycled aggregates. Fig. 2 shows the processing of recycled aggregates RCA 1, RCA 2, and RCA 3.

Table 1. Physical Properties of Fine Aggregate and Course aggregate

Aggregate	Water absorption (%)	Apparent density (kg/m^3)	Bulk density (kg/m3)	Impact value (%)	L.A. Abrasion value (%)	Specific Gravity
NA	0.4	3.0	1459	20	40.6	2.6
RCA1	5.76	2.1	1300	24	51.6	1.9
RCA2	8.68	2.16	1288.5	25	54	1.82
RCA3	9.8	2.116	1281.2	27	55.3	1.8

Fig.1 Crushing of concrete beams in Laboratory

Fig. 2 Processing of RCA 1, RCA 2, and RCA 3.

Water: It is consumable potable water that has been used. It must be free of salts and organic substances, since they may affect the qualities of fresh and hardened concrete.

Methodology: This experimental work consists of 4 cycles. 0th cycle concrete is control concrete which has ingredients like cement, fine aggregate and NCA. 1st Cycle concrete is the same as control concrete, but has the RCA which obtained from crushing of 0th cycle specimens. 2nd cycle concrete specimens were prepared using RCA which is obtained by crushing of 1st cycle specimens. 3rd cycle concrete was prepared using RCA which was obtained by crushing of 2nd cycle specimens. For 1st cycle, 2nd cycle and 3rd cycle concrete specimens, the natural aggregates were partially replaced by 25%, 50% and 100% by RCA, to assess the RAC's overall performance and to compare them with the control concrete. The concrete mix design used in this study is as per IS 10262:2009 & IS 456:2000 for M25 grade of concrete. All the mixes are designed by keeping the water content constant.
The mix proportion obtained 1: 1.81: 2.62 with a 0.45 w/c ratio. A 25 represents the 25 % replacement of NCA by RCA, A 50 represents the 50 % representation of NCA by RCA and A 100 represents the 100 % replacement of NCA by RCA.
All the specimens for the mechanical and durability test were cast and cured for period of 28 days. Mechanical properties like compressive strength were done on 150x150x150 mm cubes, split tensile strength was done on 300mm diameter and 150mm length cylinders, flexural strength was done on 500x100x100 mm beams, durability test like, water absorption test was done on 100mm diameter and a height of 50mm, abrasion test was done on 75x75x35 mm small cubes and for rapid chlorine penetration test was done on concrete cylindrical discs with a diameter of 100 mm and a height of 50 mm.

3. Results and Discussions

Compressive Strength Test: The 28-day hardened concrete cubes that were assessed using compression testing equipment in accordance with IS 516-1959, and Figure 3 depicts the average compressive strength value of three samples. By referring to figure 4, the compressive strength for NAC samples is 27.53 MPa. The compressive strength of R1 cycle concrete with A25, A50, and A100 is 25.03 MPa, 24.43 MPa, and 22.96 MPa, respectively. R2 cycle with A25, A50, and A100 is 21.27 MPa, 20.95 MPa, and 20.8 MPa, respectively. R3 cycle with A25, A50, and A100 is 20.39 MPa, 19.1 MPa, and 18 MPa, respectively. NAC has higher compressive strength than R1, R2 and R3 cycles. As the percentage of replacement of RCA increased, the compressive strength decreased substantially. The decrease in compressive strength could be attributed to poor bonding between cement pastes and RCA

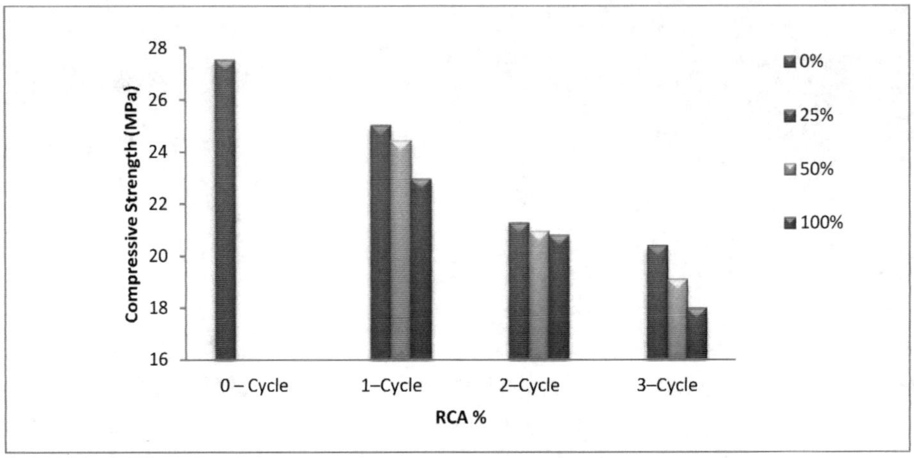

Fig. 3. Compressive strength of NAC and multiple cycles RAC

Split Tensile Strength Test: The splitting tensile strength of cylindrical specimens in UTM was performed at 28 days of normal curing in compliance with IS 5816-1999 [ZZ3]. Figure 4 shows the tensile specimen placed in UTM. Figure 6 depicts the fluctuation of the average tensile strength value of three specimens. By referring to figure 6, NAC has a tensile strength of 2.81 MPa.

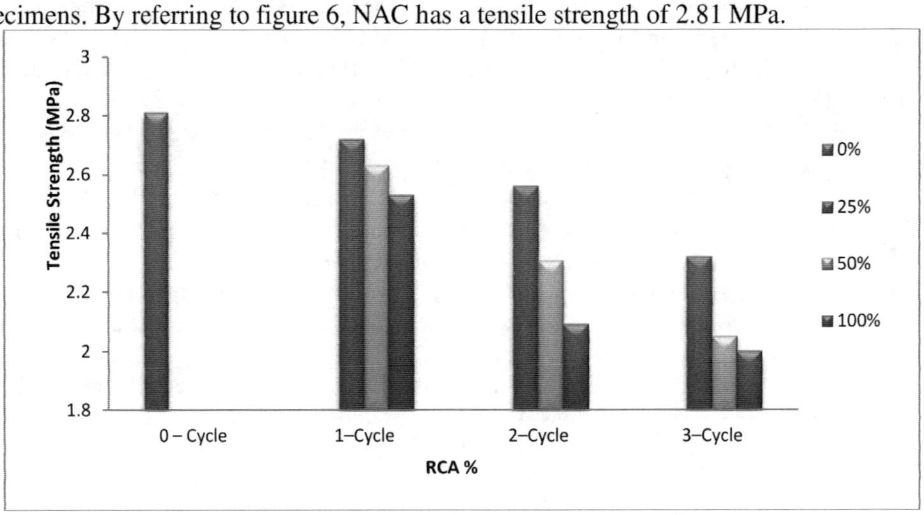

Fig. 4. Tensile strength of NAC and multiple cycle RAC

The tensile strength of R1 cycle concrete with A25, A50, and A100 were 2.72 MPa, 2.63 MPa, and 2.53 MPa, respectively. R2 cycles with A25, A50, and A100 were 2.56 MPa, 2.306 MPa, and 2.09 MPa, respectively. R3 cycle with A25, A50, and A100 was 2.32 MPa, 2.05 MPa, and 2 MPa, respectively. NAC has higher tensile strength than R1, R2 and R3 cycles. Tensile strength was reduced substantially with an increase in the number of cycles of recycling and an increase in the percentage of replacement NCA by RCA.

Flexural strength: The flexural strength of a beam specimen was measured on a UTM after 28 days in accordance with IS 516-1959. Figure 7 shows the testing of concrete prism specimens for flexural strength. Figure 5 exhibits the average flexural strength value of three specimens. In UTM, the failure load was applied under four-point loading to evaluate the flexural strength of concrete prisms. By referring to figure 8, the flexural strength of NAC was found at 4.23 MPa. The flexural strength of R1 cycle concrete with A25, A50, and A100 were 3.107 MPa, 2.935 MPa, and 2.38 MPa, respectively. R2 cycles with A25, A50, and A100 were 2.84 MPa, 2.78 MPa, and 2.325 MPa, respectively. R3 cycles with A25, A50, and A100 were 2.76 MPa, 2.59 MPa, and 1.78 MPa, respectively. NAC has higher flexural strength than R1, R2 and R3 cycles. Flexural strength was reduced substantially with an increase in the number of cycles of recycling and an increase in the percentage of replacement NCA by RCA

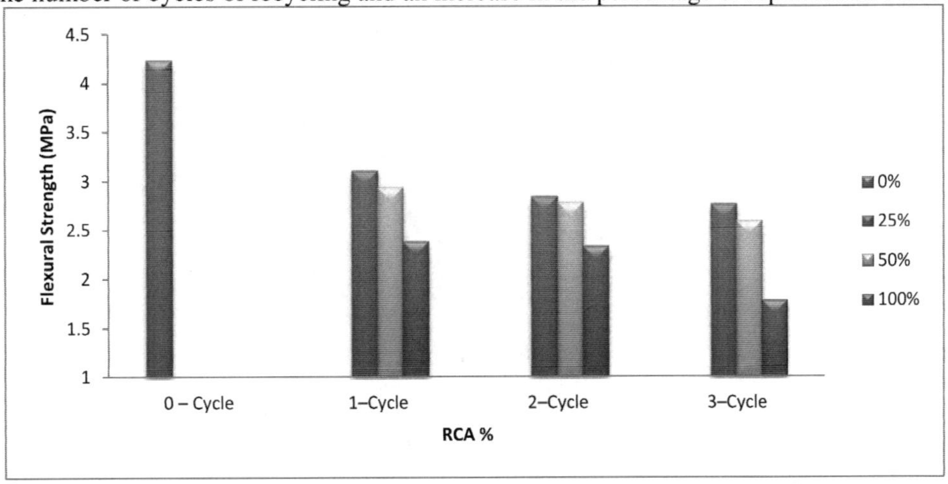

Fig. 5. Flexural strength of NAC and multiple cycle RAC

Water absorption test: Water absorption tests are performed to assess the perviousness of concrete, which is linked to its permeability. Water absorption tests were carried out on 100 mm concrete cubes in line with ASTM C642 – 2013. Figure 6 shows the water-immersed cube specimens during their 28 days curing period. After the curing period, the samples were dried in the open air for 2–3 hours to remove any remaining surface moisture. The samples were now stored in a 110°C oven. The samples were taken out of the oven after 24 hours and allowed to cool to room temperature before being used. Referring the Figure 6, the water absorption of NAC was found at 3 %. The water absorption of R1 cycle concrete with A25, A50, and A100 were 3.52 %, 4.2 %, and 4.4 % respectively. R2 cycle with A25, A50, and A100 were 5.03 %, 5.4 % and 6.1 % respectively. R3 cycle with A25, A50, and A100 were 6.33 %, 6.55 % and 6.91 % respectively. NAC has lower water absorption than R1, R2 and R3 cycles. Water absorption was increased substantially with an increase in the number of cycles of recycling and an increase in the percentage of replacement NCA by RCA. According to the findings, RAC cubes absorb more water than NAC cubes. The excessive water absorption is generated by mortar sticking to the RCA surfaces, hence the water absorption characteristic of RAC increases with the number of recycling cycles and the percentage of replacing NCA by RCA.

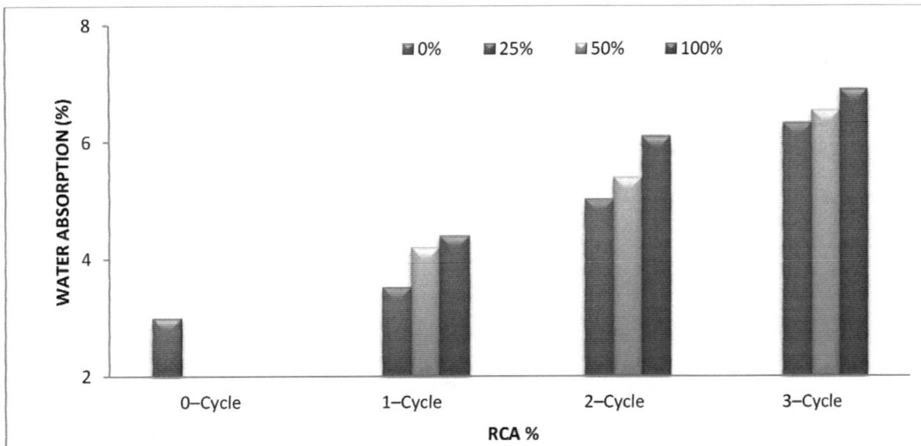

Fig. 6. Water absorption of NAC and multiple cycle RAC

Rapid chloride penetration test: The rapid chloride permeability test was carried out in accordance with ASTM C1202-2009 [MM], with the setup indicated in figure 7.

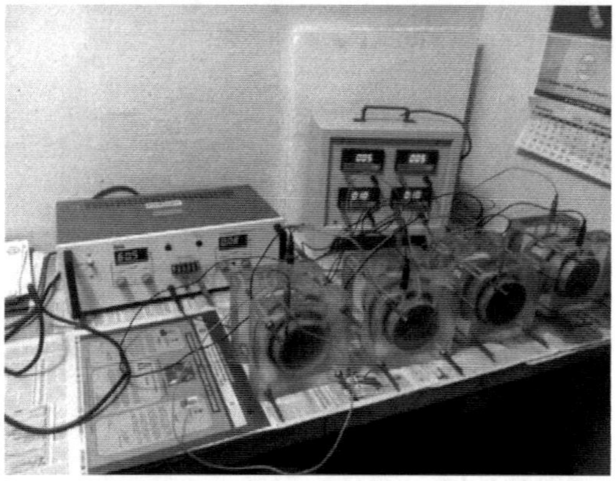

Fig.7. Rapid chloride penetration test Oven

Concrete cylindrical discs of 100 mm diameter and 50 mm height were formed and dried for 28 days. The disc was then filled with 0.3 per cent NaOH on one end and 3 per cent NaCl on the other. A 60 V potential difference between the two cells was maintained for 6 hours, and the charge passed through the material was measured. Coulombs were used to measure chloride penetration (C). Figure 8 shows the RCPT for NAC specimens at 527.1 C.

The RCPT of R1 cycle concrete with A25, A50, and A100 were 702.3 C, 858.3 C and 942.3 C respectively. R2 cycle with A25, A50, and A100 were 1349.4 C, 1424.7C and 1430.2 C respectively. R3 cycle with A25, A50, and A100 was 1625.1 C, 1767.9 C, and 3065.7 C respectively. RCPT increases as the fraction of RCA replacement increases and also with the multiple cycles of recycling the RCA. The higher the conductivity, the greater the chance of penetration and easy corrosion. It means that the RAC was less corrosion resistant, and less durable.

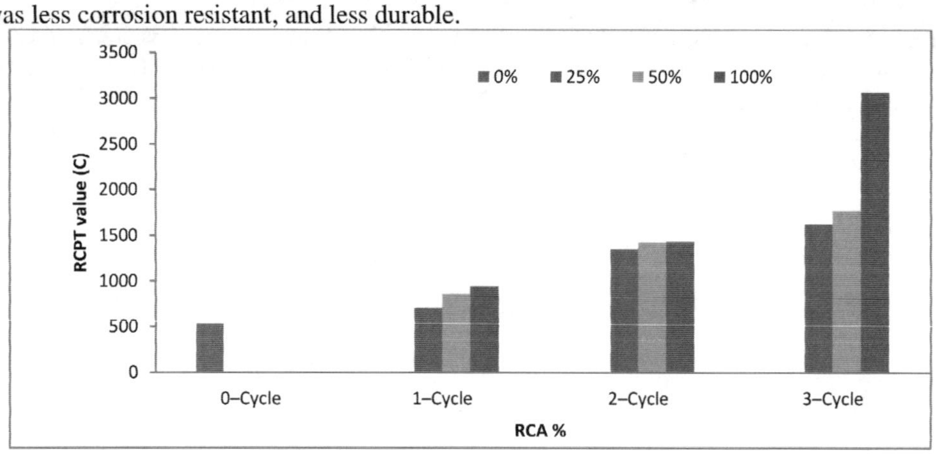

Fig. 8. RCPT value for RCA cycles

Abrasion test: After casting and curing for 28 days, an abrasion test was performed in accordance with IS 1237:1980 on a square test specimen of (70X70X35mm), with the abrasion powder utilised having an aluminium oxide 80 grade content of not less than 95 % by mass. Specimens were dried for 24 hours prior to testing and their dry weight was recorded. On the upper surface of the specimen, five spots are marked: four in the corners and one in the centre. The thicknesses of the samples were measured at these spots with digital callipers prior to being inserted in the socket of the abrasion testing equipment

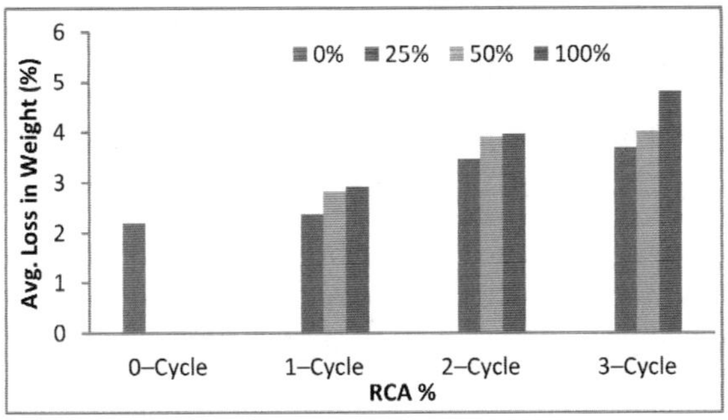

Fig. 9. Abrasion percentage loss in weight

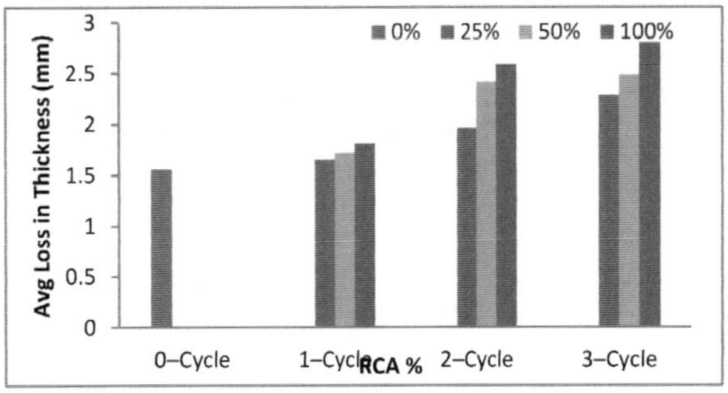

Fig. 10. Loss in thickness in abrasion test

The grinding disc is configured to rotate at a speed of 30 revolutions per minute. The abrasive charges are fed through a funnel provided in the setup. After every 22 revolutions, the specimen was rotated about its vertical axis at an angle of 90° in a clockwise direction. The preceding technique is repeated 16 times for a total of 22 revolutions. The weight and thickness of the specimen are measured at the same 5 places after each cycle. Referring to Figure 9, the loss of weight for NAC specimens is 2.2 %. R1 cycle concrete with A25, A50, and A100 were 2.38 %, 2.82 %, 2.92 % respectively. R2 cycle with A25, A50, and A100 were 3.7 %, 3.9 % and 3.97 % respectively. R3 cycle with A25, A50, and A100 were 3.69 %, 4.01 % and 4.81 % respectively. Loss of weight in percentage increases as the percentage of RCA replacement increases and also with the multiple cycles of recycling the RCA. Referring to Figure 10, the average loss in thickness for NAC specimens is 1.56 mm. R1 cycle concrete with A25, A50, and A100 was 1.65 mm, 1.71 mm and 1.81 mm respectively. R2 cycles with A25, A50, and A100 were 1.96 mm, 2.41 mm and 2.59 mm respectively. R3 cycles with A25, A50, and A100 were 2.28 mm, 2.48 mm and 2.8 mm respectively. Average loss in thickness increases as the percentage of RCA replacement increases and also with the multiple cycles of recycling the RCA.

4. Conclusion

Compressive strength, split tensile strength, and modulus of rupture all drop as the number of cycle's increases. Concrete's water absorption and abrasion resistance increase as the number of cycles increases (RCA1, RCA2 & RCA3). And also it is seen that with an increase in the %age replacement of RCA (25%, 50% & 100%) water absorption and abrasion values also increase. The resistance to the chloride ion penetration is more in the conventional concrete when compared to that of the RCA concrete. From the results, it can be seen that chloride ion penetrability for all RCA concrete varies from very low to moderate. From the existing studies, it can be concluded that 25% RCA1 is suitable for construction purposes since the results obtained are almost equal to that of conventional concrete.

5. References

[1] S.K. Kirthika, S.K. Singh, A. Chourasia, Alternative Fine Aggregates in Production of Sustainable, J. Clean. Prod. (2020) 122089.

[2] A. Akhtar, A.K. Sarmah, Construction and demolition waste generation and properties of recycled aggregate concrete: A global perspective, J. Clean. Prod. (2018).

[3] W. Alnahhal, O. Aljidda, Flexural behavior of basalt fiber reinforced concrete beams with recycled concrete coarse aggregates, Constr. Build. Mater. 169 (2018) 165–178.

[4] N. Garcia-troncoso, L. Li, Q. Cheng, K. Hung, T. Ling, Case Studies in Construction Materials Comparative study on the properties and high temperature resistance of self-compacting concrete with various types of recycled aggregates, Case Stud. Constr. Mater. 15 (2021) e00678.

[5] B. Jaivignesh, A. Sofi, A study on the performance of concrete containing recycled aggregates and ceramic as materials replacement.

[6] A. Mar, J.A. Mesa, Mechanical Properties of Concrete Using Recycled Aggregates Obtained from Old Paving Stones, (2021) 1–17.

[7] P. Tamayo, J. Pacheco, C. Thomas, J. De Brito, J. Rico, applied sciences Mechanical and Durability Properties of Concrete with Coarse Recycled Aggregate Produced with Electric Arc Furnace Slag Concrete, (2019).

[8] V.W.Y. Tam, M. Soomro, A. Catarina, J. Evangelista, A review of recycled aggregate in concrete applications (2000 – 2017), Constr. Build. Mater. 172 (2018) 272–292.

[9] M. Arezoumandi, A. Smith, J.S. Volz, K.H. Khayat, An experimental study on flexural strength of reinforced concrete beams with 100 % recycled concrete aggregate, Eng. Struct. 88 (2015) 154–162.

[10] J.F. Dong, Q.Y. Wang, Z.W. Guan, Material properties of basalt fibre reinforced concrete made with recycled earthquake waste, Constr. Build. Mater. 130 (2017) 241–251.

[11] G. Dimitriou, P. Savva, M.F. Petrou, Enhancing mechanical and durability properties of recycled aggregate concrete, Constr. Build. Mater. 158 (2018) 228–235.

[12] B. Ali, L.A. Qureshi, Influence of glass fibers on mechanical and durability performance of concrete with recycled aggregates, Constr. Build. Mater. 228 (2019) 116783.

[13] N. Jain, M. Garg, A.K. Minocha, Green Concrete from Sustainable Recycled Coarse Aggregates : Mechanical and Durability Properties, 2015 (2015).

[14] M. Surendar, G.B. Gnana, M. Sharaniya, M.S. Deepak, T. V Soundarya, Mechanical properties of concrete with recycled aggregate and M – sand Materials Today : Proceedings Mechanical properties of concrete with recycled aggregate and M À sand, Mater. Today Proc. (2021).

[15] I. Nováková, K. Mikulica, Properties of concrete with partial replacement of natural aggregate by recycled concrete aggregates from precast production, Procedia Eng. 151 (2016) 360–367.

[16] Y. Zhao, W. Zeng, H. Zhang, Properties of recycled aggregate concrete with different water control methods, Constr. Build. Mater. 152 (2017) 539–546.

[17] L. Li, D. Xuan, A.O. Sojobi, S. Liu, S.H. Chu, C.S. Poon, Development of nano-silica treatment methods to enhance recycled aggregate concrete, Cem. Concr. Compos. 118 (2021) 103963.

[18] P. Pliya, H. Hajiloo, S. Romagnosi, D. Cree, S. Sarhat, M.F. Green, The compressive behaviour of natural and recycled aggregate concrete during and after exposure to elevated temperatures, J. Build. Eng. 38 (2021) 102214.

[19] A. Singh, Z. Duan, J. Xiao, Q. Liu, Incorporating recycled aggregates in self- compacting concrete : a review, J. Sustain. Cem. Mater. 9 (2020) 165–189.

[20] C. Laneyrie, A. Beaucour, M.F. Green, R.L. Hebert, B. Ledesert, A. Noumowe, Influence of recycled coarse aggregates on normal and high performance concrete subjected to elevated temperatures, 111 (2016) 368–378.

Study on International Road Roughness index (IRI) using Smart phone application from REVA University to Kodigehalli gate, Bangalore, India

Dr P. Shivananda[1*], Sangram Kumar Khatua [2]

[1]Professor, School of civil engineering, REVA University, Yelahanka, Bengaluru-560064
[2]PG Student, School of civil engineering, REVA University, Yelahanka, Bengaluru-560064
*(p.shivananda@reva.edu.in, sangramkumar71@gmail.com)

Abstract. International Roughness Index (IRI) is one of the most widely used indices to determine the roughness of different road surface. Roughness index helps in identifying the quality of the road surface and suggest improvement measures. In this paper Roughness index values are measured from REVA University to Kodigehalli Gate. International Roughness Index are measured using mobile application. IRI value depends on vehicle speed and road. Roughness indicates the present road condition and helps in making decision related to management and maintenance of different types of roads. Based on Carbin app [1] results road user can identify 1) Safety, 2) Traffic, 3) Comfort of driving, 4) Fuel Consumption, 5) Quality of Air and 6) Road condition. Carbin application is used to collect Road quality, Speed, CO2 savings. In the study stretch Maximum value of IRI = 5.6m/km obtained from Bagalur cross to on BSF Cross and this stretch road user face all the above 6 problems. Minimum value of IRI= 2.44m/km obtained from Kodigehalli Cross to Bagalur cross (Airport Road-NH), the road is recently maintained by overlaying, in this stretch, road users comfortably use this stretch of road with respect above 6 aspects. As per the IRI value, new roads to be constructed to get IRI<1.2 m/km, once it reaches IRI=4 m/km road need to be maintained to get IRI < 2.2 m/km.

Keywords: Carbin app, International Roughness index (IRI), Road quality.

1. Introduction

Road networks are expanding all over the world but in the emerging economies their growth is at a historic high. From S. Fan and C. Chan-Kang[2]only 147 km of expressway in 1989, China has added more than 1 million km . From, Ministry of road transport and highways [3],India has built nearly 100,000 km of roads in the last 15 years. With more addition of new roads. Maintenance of the existing road making a bigger challenge. Deterioration of due to harsh environment, use of low-quality materials and non-technical construction, and overloading of traffic. Bangalore Being a metropolitan city has high density of living environment and active economic activities. Urban roads are planned as interconnected network for easy movement of traffic. Transportation engineers must give at most priority for safety of road users. Efforts should be made in order to find out quality of road which indirectly or directly reflects safety of the road users. Travel behavior depends on connectivity and quality of road[4]. Many researchers V. Khalifeh and at al [5], P. Múčka [6], P. Múčka[7] and M. R. De Blasiis and at al [8], have worked on International Roughness Index and IRI as used a yardstick to check the quality of Road. Vehicles are equipped with GPS tools are used to track the movement of vehicle[9]. Now a days smartphone are used find IRI values [10] [11]. determining. In present study cloud based Carbin app is a GPS based navigation app used to determine IRI, CO2 Savings, distance for any stretch[1]. Speed-related IRI thresholds at different speeds are classified as Very Good, Good, Fair, Mediocre and Poor[12]. Google map gives the dynamic updates about the traffic and gives the suggestions of alternate route for avoiding the traffic jam [13], similarly IRI value with colour code helps in giving single to maintain the road for both road users and maintenance department. The traditionally followed operational plan for the public transportation system in urban areas results in a financial loss of 7% [9], by providing integrated transport system(BMTC to Metro) helps to reduce economical loss. Bangalore is one of the fast-growing IT hubs has traffic scourge with rising vehicle ownership, increased traffic congestion results in huge economic loss due to traffic congestion where fuel loss is estimated at Rs 2.8 lakh litres per hour, 50 crore litres per year, in monetary terms this translates to a loss of Rs 3700 crores per year and Rs 2350 crores towards work hours lost, it has been observed that the city is in the grip of rising levels of concentration of pollutants through vehicle emission [14], by identifying actual IRI value of roads and maintaining roads to lower IRI helps in reducing fuel, time and CO2 emission.

2. Objectives of the study

2.1. To determine

2.1.1. Cabin Result

2.1.2. Road Quality (IRI)

2.1.3. CO2 Savings

2.1.4. Time

2.1.5. Distance travelled for selected stretch of road using Carbin app.

2.2. Based on IRI value roads stretches are classified accordingly and identified necessary maintenance.

3. Study area

Selected study area from REVA University to Kodigehalli Gate consist of total length 10.9km. Includes NH44, SH 104 and Urban Road. This road provide services as NH, SH, Education institutes, Air port, SEZ, Hospitals, Commercial peace and Residential area. Hence road quality studies are conducted.

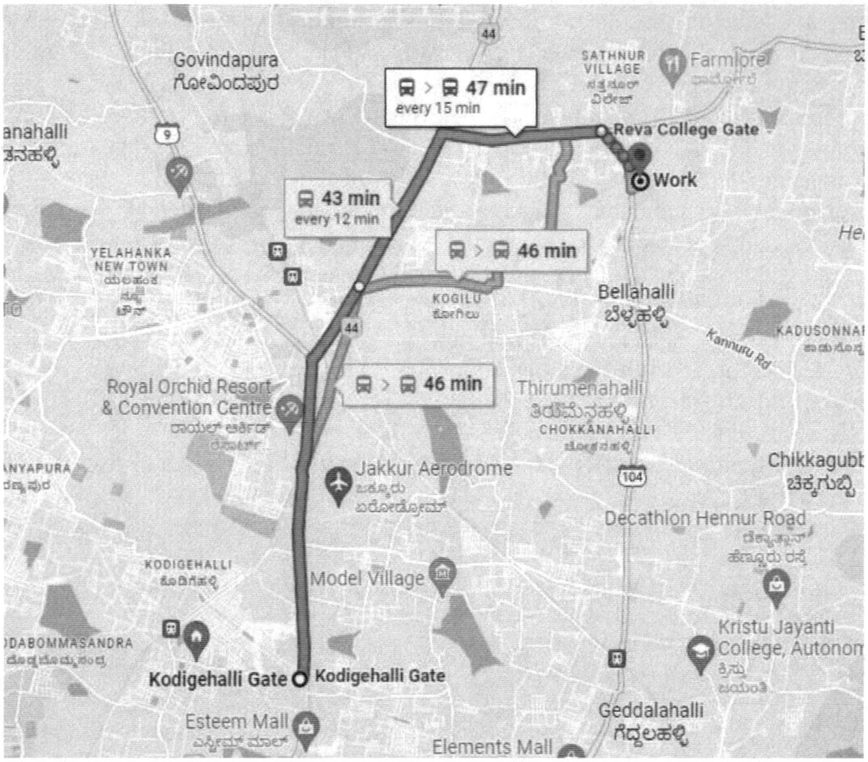

Fig.1: Route map of study area. (Source: Google maps)

4. Methodology:

4.1. Carbin App can be used with following settings shown in Table 1.

Table1 Carbin App Setting in App

Carbin App Setting			
Sl No	Details	Choice for Selection	Selected choice for present study
1	Distance	mi/km/m	km
2	Fuel	Gasoline/Diesel	Gasoline
3	IRI Unit	in/mi or m/km	m/km
4	CO2 Unit	lbs/kg	kg
5	Vehicle type	Car/Truck	Car

4.1.1. Vehicle used - Car
4.1.2 Smart phone- Keeping smart phone in firm potion on dash board such that Carbin value >80%
4.1.3 App Used- Carbin App downloaded using Google app store.
A new app called "Carbin" allows users to crowdsource road-quality data, particularly road roughness, with their smartphones. Using the accelerometers found in our everyday mobile devices, Carbin converts vehicle acceleration signals into standard measurements IRI used by most Departments of Transportation. Carbin Application is a GPS based application where the road quality data's are stored in the interface and same data's are utilised in order to study the quality of pavement. the parameter which can be obtained using this application are CO_2 savings, IRI (road quality), Distance and Speed.

5. Results and Discussion

5.1. Collection of data:

Data's are collected using Carbin app from REVA college to Kodigehalli Gate. Strech is divided in to three division.

5.1.1. Kodigehalli Gate to Bagalur Cross -NH 4 (Resurfaced Good Bitumen Road) as shown in Fig 2.

5.1.2. Bagalur Cross to BSF Circle-Urban Road (Unmaintained Bitumen Road) as shown in Fig 3.

5.1.3. BSF Cross to REVA University – SH 104 (Resent White toped Road) as shown in Fig 4.

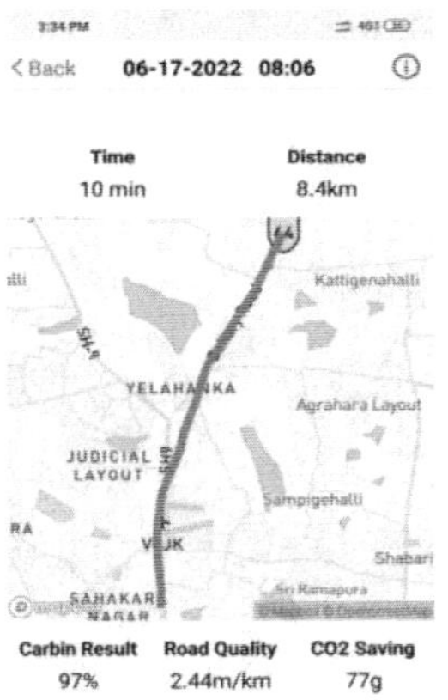

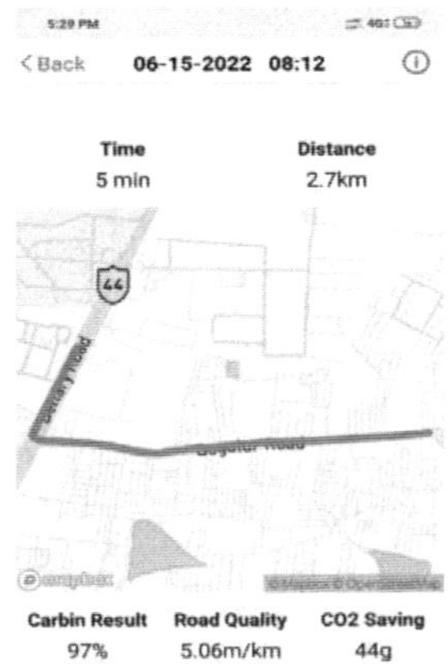

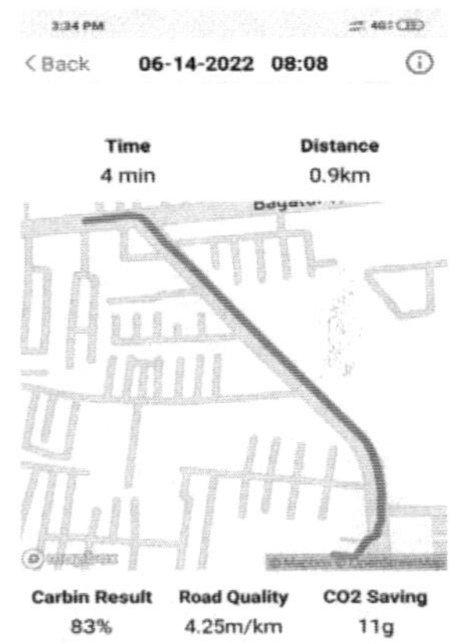

Fig 2 Kodigehalli Gate to Bagalur;. Fig 3 Bagalur Cross to BSF Circle; Fig 4 BSF Cross to REVA;

Table 2. Data obtained from Carbin app for three stretches.

Stretch	From -To	Carbin results	Data from Carbin app				Calulate Values	
			Road quality (IRI in m/km)	CO2 savings in g	Distance in km.	Time in min	Speed km/hr	CO2 savings in g /min
1	Kodigehalli Gate to Bagalur Cross (NH)	97%	2.44	77	8.4	10	50.4	0.8
2	Bagalur Cross to BSF Cross (UR)	97%	5.6	44	2.7	5	32.4	0.5
3	BSF Cross to REVA University.(SH)	83%	4.25	11	0.9	4	13.5	0.2

Table 3. Speed-related IRI thresholds at different speeds suggested S. L. Chen[12]

Ride Quality Level	IRI Thresholds at Different Speeds (m/km)					
	20	40	60	80	100	120
Very Good	<5.72	<2.86	<1.90	<1.43	<1.14	<0.95
Good	5.72-8.99	2.86-4.49	1.90-2.99	1.43-2.24	1.14-1.79	0.95-1.49
Fair	9.00-11.39	4.50-5.69	3.00-3.79	2.25-2.84	1.80-2.27	1.50-1.89

Mediocre	11.40-16.16	5.70-8.08	3.80-5.40	2.85-4.05	2.28-3.24	1.90-2.70
Poor	>16.16	>8.08	>5.40	>4.05	>3.24	>2.70

5.2. Riding Quality level

By comparing Data from Table 2 and IRI threshold from Table 3

5.2.1. For Stretch 1, Road is NH speed allowed is 60-80 kmph IRI value obtained is 2.44m/km from Table 2, from Table 3, Road is Fair to Good.

5.2.2. For Stretch 2, Road is Urban Road speed allowed is 40-60 kmph IRI value obtained is 5.6m/km from Table 2, from Table 3, Road is Poor to Fair.

5.2.3. For Stretch 3, Road is SH Road speed allowed is 40-60 kmph IRI value obtained is 4.25m/km from Table 2, from Table 3, Road is Good to Mediocre.

5.3. Carbon Dioxide Savings.
From Carbon dioxide savings data from Table 1, Stretch 1 is 0.8g/min is highest compared to Stretch 2 is 0.5g/min and stretch 3 is 0.2 g/min. Hence Stretch 1 is good as per Carbon dioxide savings

6. Conclusions
Conclusions based on the IRI study carried from REVA University to Kodigehalli Gate by using Carbin app are

1. Roughness index was ranging from 2.44 m/km to 5.6m/km, IRI data obtained indicates Poor to Good. Hence roads need maintenance and relaying and to bring IRI value less than 2m/Km.
2. CO_2 savings was found minimum of 0.2g/min to maximum of 0.8 g/min, by maintaining IRI value less than 2m/Km, CO_2 savings increases to 3g/Min.
3. By observing the IRI value using Carbin app helps to decide maintenance requirement of the road. Proper maintenance reduces fuel consumption, damage of vehicle and CO_2 emission.

7. Acknowledgement

Authors would like to acknowledge the Carbin App developers and MIT CS Hub.

8. References
[1] A. Us, C. Science, and E. Carbon, "Crowdsourced Pavement Data : The Carbin App Measuring Road Quality and Excess Fuel Consumption wit … The Carbin App : Crowdsourcing Pavement Data," no. July 2020, pp. 2020–2022, 2021.
[2] S. Fan and C. Chan-Kang, "Road development, economic growth, and poverty reduction in China," *Res. Rep. Int. Food Policy Res. Inst.*, no. 138, p. 138, 2005, doi: 10.2499/0896291413rr138.
[3] Ministry of road transport and highways, "Annual report 2021-2022," 2022.
[4] P. Shivananda, "Travel Behaviour of REVA University Students and Staff Literature ". July, 2020.
[5] V. Khalifeh, A. Golroo, and K. Ovaici, "Application of an Inexpensive Sensor in Calculating the International Roughness Index," *J. Comput. Civ. Eng.*, vol. 32, no. 4, 2018, doi: 10.1061/(asce)cp.1943-5487.0000761.
[6] P. Múčka, "Influence of Profile Specification on International Roughness Index," *J. Infrastruct. Syst.*, vol. 25, no. 2, p. 04019005, 2019, doi: 10.1061/(asce)is.1943-

555x.0000478.
[7] P. Múčka, "Road Roughness Limit Values Based on Measured Vehicle Vibration," *J. Infrastruct. Syst.*, vol. 23, no. 2, p. 04016029, 2017, doi: 10.1061/(asce)is.1943-555x.0000325.
[8] M. R. De Blasiis, A. Di Benedetto, M. Fiani, and M. Garozzo, "Assessing of the road pavement roughness by means of lidar technology," *Coatings*, vol. 11, no. 1, pp. 1–19, 2021, doi: 10.3390/coatings11010017.
[9] C. Prathyusha, S. Singh, and P. Shivananda, "Strategies for Sustainable, Efficient, and Economic Integration of Public Transportation Systems BT - Urban Science and Engineering," 2021, pp. 157–169.
[10] R. Kumar, A. Mukherjee, and V. P. Singh, "Community Sensor Network for Monitoring Road Roughness Using Smartphones," *J. Comput. Civ. Eng.*, vol. 31, no. 3, p. 04016059, 2017, doi: 10.1061/(asce)cp.1943-5487.0000624.
[11] P. Kamranfar, D. Lattanzi, A. Shehu, and S. Stoffels, "Pavement Distress Recognition via Wavelet-Based Clustering of Smartphone Accelerometer Data," *J. Comput. Civ. Eng.*, vol. 36, no. 4, pp. 1–12, 2022, doi: 10.1061/(asce)cp.1943-5487.0001022.
[12] S. L. Chen, C. H. Lin, C. W. Tang, L. P. Chu, and C. K. Cheng, "Research on the international roughness index threshold of road rehabilitation in metropolitan areas: A case study in Taipei city," *Sustain.*, vol. 12, no. 24, pp. 1–19, 2020, doi: 10.3390/su122410536.
[13] M. P. S. K. P.Shivananda, "Google Map Usage in Transportation System," vol. 13, no. 4, pp. 3085–3096, 2020.
[14] P. Challa and P. Shivananda, "A Review on Integrated Transportation System for," vol. 5, no. X, pp. 1260–1268, 2017.

A Laboratory evaluation of Paper mill waste in manufacturing of Clay Bricks

Raveesh J[1], Vishnu R Uppin[2],Chandan M R[3] , Dr. Nagaraj Gupta[4]
1,2,3,4 Assistant Professor, School of Civil Engineering, REVA University, Bengaluru – 560064, Karnataka, India.
Assistant Professor, Dept of Civil Engineering, Gopalan college of Engineering and management, Bengaluru – 560048, Karnataka, India.
Professor, Dept of Civil Engineering, EWIT,Bengaluru – 560091, Karnataka, India.
(raveeshj27@gmail.com ; chandugowda26@gmail.com, nagarajaguptams@gmail.com)

Abstract. India has one of the world's fastest expanding pulp and paper markets. With a growth rate of more than 10% per year in terms of per-capita growth, The pulp and paper industry requires a lot of capital investment, energy, and water. A paper mill produces 40 to 50 kg of dry sludge (70 percent primary sludge and 30 percent secondary sludge) when 1 tonne of paper is produced. Most of this sludge is now disposed of by land spreading and filling, both of which involve high disposal costs. Pulp and paper industry is one of the most polluted industries in the world in general and India in particular. India produces 10.11 million tons paper per annum which is 2.6% of world's overall production quantity of paper. This study present using paper mill waste as a component in the manufacture of burnt bricks as a possible route to environmentally friendly construction. This research report details the findings of an experimental investigation into the production of eco-friendly, bricks using soil and paper mills. This research exhibits some increased compressive strength in mix of papermill waste.

Keywords: Paper mill, Clay Brick, Compression Testing Machine, Laterite soil

1. Introduction

One of the world's oldest building materials is clay brick. The majority of buildings in the world are composed of clay bricks, and they continue to provide outstanding service. Before concrete and other modern construction materials hit the market, bricks were a common building material. The productive re-use of waste material represents a way of solving the major concern of solid waste management (Davis & Cornweel, 1998).Because they use natural clay elements, bricks are a popular construction material in Asia. In India, 250 billion bricks are produced annually. After China, it is the largest country in the world. Brick construction is a common choice in the Asian region due to its inexpensive cost of construction and simplicity of manufacture using just earth. India is the world's second-largest manufacturer of bricks, behind China, with an annual production of roughly 250 billion. One of the key materials utilized in the construction business is clay brick. Bricks have been left behind in advancement by the traditional technique of production. The ecology is impacted by the growing population and increased construction activities. The demand for research into how this can be accomplished while benefiting the environment and maintaining the material requirements affirmed in the standards has been prompted by the rise in popularity of using environmentally friendly, inexpensive, and lightweight construction materials in the building industry. Due to its characteristics, brick is one of the masonry units that may be used in construction the most easily. Attempts have been made to manufacture low-cost, environmentally friendly bricks using waste materials. Traditional structures made of heavy bricks required a strong foundation, which significantly increased the construction cost. With new advances in construction, brick walls no longer carry structural loads, and thus performance requirements such as strength are not as stringent as they once were. Furthermore, the most significant advancement in the history of brick masonry construction has been to improve their thermal insulation properties and high specific weight. This is related to the fact that brick manufacturing consumes a significant amount of fertile agricultural soil.

The main objective of this study is to The major goal is to utilize some paper mill waste when building bricks and reduce the amount of fly ash in the finished product. To ascertain the physical properties of the soil material. Pulp manufacturing starts with raw material preparation, which includes debarking, chipping, chip screening, chip handling, and storage (Bajpai 2011, Brännvall 2009).

To analyse soil material with a pulp and paper mill to identify its physical qualities. To evaluate the mechanical toughness of bricks made from waste from paper mills. To assess the brick's financial viability

2. Materials and Methodology

2.1 Materials
The materials used for present study is Laterite soil, Fly ash and Paper mill waste.
2.1.1 Soil:
For this study, combination of two different types of dirt alluvial soil and lacustrine soil for the brick's construction. Bricks were produced using a soil mixture of lacustrine and alluvial origin.

Figure 1. Soil Sample

2.1.2 Fly ash:
Flyash is a finely separated byproduct of pulverised coal combustion that is carried away from the combustion chamber by exhaust gases. In 2021, more than 61 million metric tonnes (68 million tonnes) of fly ash were generated.

Figure 2. Fly ash

2.1.3 Paper Mill Waste and Pulp:
Waste paper recycling which is involves mixing used paper with water and chemicals to break it down. Waste paper is broken down through the recycling process by being combined with water and chemicals.

The resulting mixture, known as pulp, is next diced and burned, further breaking it down into cellulose strands, a type of organic material.

Figure 3. Paper mill waste

2.2 Methodology

Methodology which comprise the tests on Soil materials, tests on paper mill and Tests on Bricks are tabulated in table 1, 2, 3&4.

3. Results and Discussions

The results obtained in the experimental investigation are as fallows.

Table 1. Tests on soil

Sl no.	Test conducted	Results	IS code Requirements
1.	Specific gravity	2.82	2.67-2.9
2.	Moisture content	18.23%	15%-20%
3.	Liquid limit	43.52	40-50
4.	Plastic limit	22.35	18-25
5.	Maximum dry density	17.8	-
5.	Plasticity index	21.17	20-35
6.	CBR Test	9.2%	-

Table 2. Physical properties of paper sludge ash

Sl no.	Test conducted	Results
1.	Specific gravity	2.45
2.	Colour	Greyish white

Table 3. Tests on brick

Sl no.	Test conducted	Results
1.	Hardness test	This test showed that the brick containing 10% to 25% off paper pulp was hard enough to withstand the scratches.
2.	Soundness test	All the bricks emitted a good ringing sound. As per I.S standards
3.	Weighment test	The brick containing 10% addition of paper sludge and fly ash weighed 450grams lesser than traditional brick. The brick containing 15% addition of paper sludge and 10% of fly ash weighed 550 grams lesser than traditional brick. The brick containing 20% addition of paper sludge and 10% of fly ash weighed 520 grams lesser than traditional brick.
5.	Efflorescence test IS 3495 (part 3)	The test showed satisfactory result
6.	Compressive test IS 3495 (part 1)	The table number below represents compressive test results.

Table 4. Compression strength on bricks

Sl no.	Percentage of vary with paper mill waste sludge	Compression test value in(kg/cm^2)
1	0%	107
2	5%	112
3	10%	117
4	15%	95
5	20%	75

Table 5. Compression strength on bricks vary with Fly ash with 10% of Paper mill waste

Sl no.	Percentage of vary with fly ash	Compression test value in(kg/cm^2)
1	0%	107
2	10%	120
3	20%	118

Table 6. Comparison of cost (Indian Rupee) with conventional bricks (Manufacturing cost for 1 bricks)

Items	Conventional Bricks	Paper mill waste bricks
Materials (Kg)	2.25/brick	1.95 /brick
Salary, Mould charge, electricity, Maintenance etc.	2.35/Brick	2.35/Brick
Total	Rs:-4.6	Rs:-4.3

4. Conclusion

This paper presents use of paper mill waste in some extent as admixtures with soil in manufacturing of clay bricks, laboratory tests have been conducted on hand mould bricks composed of varied proportions. Based on study following conclusions are drawn:

The compressive strength of Bricks increases with 10% of paper mill waste and further increased in mixes shows reduction in strength. In the present study shows that compressive strength increases with vary with fly ash content upto 20%.

In this present study, we compared with cost of manufacturing bricks by use of paper mill waste with conventional bricks, which shows to reduce cost for manufacturing bricks.

5. References

[1] Raju Sarkar, Ritesh Kurar, Ashok Kumar Gupta, Ankur Mudgal & Varun Gupta 2017, Use of paper mill waste for brick making Sarkar et al., *Cogent Engineering (2017),* Vol 4 pp.1405768,
[2] Malhotra, S. K., & Tehri, S. P. (1996). Development of bricks from granulated blast furnace slag. *Construction and Building Materials,*Vol *10(3),* pp. 191–193.
[3] Modolo, R., Benta, A., Ferreira, V. M., & Machado, L. M. (2010). Pulp and paper plant wastes valorisation in bituminous mixes. *Waste Management, 30(4),* pp.685–696.
[4] Bajpai, P. (2014). Generation of waste in pulp and paper mills. Management of pulp and paper mill waste (1st ed.). NewYork, NY: Springer.
[5] Amrilphale, S. S. and Patel, M., Utilization of red mud, fly ash for manufacturing bricks with pyrophyllite. Silic. Ind., 1987, 52(3-4), pp.31–35.
[6] Indian Standard: 2720 - (Part V). (1985). Methods of test for soils: Determination of liquid and plastic limit. New Delhi:*Bureau of Indian Standards.*
[7] Indian Standard: 2720 - (Part VII). (1980). Methods of test for soils: Determination of water content-dry density relation using light compaction. New Delhi: *Bureau of Indian Standards.*
[8] Indian Standard: 3495 (Part 1). (1992). Methods of tests of burnt clay building bricks: Part 1: Determination of compressive strength. New Delhi: *Bureau of Indian Standards.*
[9] Indian Standard: 3495 (Part 2). (1992). Methods of tests of burnt clay building bricks: Part 2: Determination of water absorption. New Delhi: *Bureau of Indian Standards.*
[10] Bajpai P (2011) *Environmentally-friendly production of pulp and paper. Wiley*
[11] Davis, M. L., & Cornweel, D. A. (1998). Introduction to environmental engineering (3rd ed.). New York, NY: *WCB McGraw-Hill.*

AUTHOR INDEX